澜湄职业教育培训中心暨柬埔寨鲁班工坊系列教材
A Series of Textbooks for Lancang-Mekong Vocational Education Training Center and Cambodia Luban Workshop

钳工技能实训教程（初级）
Training Course for Benchwork (Elementary)

主　编　米　晶
Chief editor　MI Jing
副主编　宋全胜　任慧芳　田佳峰
Deputy editor　SONG Quansheng　REN Huifang　TIAN Jiafeng

西安电子科技大学出版社

Brief Introduction

This book is compiled in accordance with the requirements of *National Vocational Skill Standard for Fitters* (*Elementary*) in terms of primary theoretical knowledge and related skills and based on the actual situation of Lancang-Mekong Vocational Education Training Center. The book mainly includes basic knowledge of fitter, parts processing operation, simple mechanical assembly, mechanical equipment testing and debugging, etc. As a supporting textbook for Lancang-Mekong Vocational Education Training Center, this book is designed bilingually in both Chinese and English for domestic readers who are seeking for professional English as well as overseas readers for their skills training or self-study.

This book can be used as a textbook for mechatronics, numerical control technology and other related majors in application-oriented undergraduate or higher vocational colleges, and can also be used as a reference book or skill training book for related technical personnel.

图书在版编目(CIP)数据

钳工技能实训教程：初级 = Training Course for Benchwork(Elementary)：英文 / 米晶主编. -- 西安：西安电子科技大学出版社，2024.2
ISBN 978-7-5606-7146-8

Ⅰ. ①钳… Ⅱ. ①米… Ⅲ. ①钳工—高等学校—教材—英文 Ⅳ. ①TG9

中国国家版本馆 CIP 数据核字(2024)第 007996 号

策　　划　秦志峰
责任编辑　秦志峰
出版发行　西安电子科技大学出版社（西安市太白南路 2 号）
电　　话　(029)88202421　88201467　邮　　编　710071
网　　址　www.xduph.com　电子邮箱　xdupfxb001@163.com
经　　销　新华书店
印刷单位　西安日报社印务中心
版　　次　2024 年 2 月第 1 版　2024 年 2 月第 1 次印刷
开　　本　787 毫米×1092 毫米　1/16　印张 10
字　　数　231 千字
定　　价　36.00 元
ISBN 978-7-5606-7146-8 / TG
XDUP 7448001-1

General Foreword

Serving"the Belt and RoadInitiative" of China, the Lancang-Mekong Vocational Education Training Center and Cambodia Luban Workshop is a joint project undertaken by Tianjin Sino-German University of Applied Sciences(TSGUAS) for the Ministry of Foreign Affairs of the People's Republic of China, the Ministry of Education and the Tianjin Municipal People's Government. Based in Cambodia, the project is designed to serve five countries in the Lancang-Mekong area and radiate to other ten ASEAN countries. It integrates functions of vocational training, vocational education, scientific research, cultural inheritance and innovation & entrepreneurship, develops both academic and non-academic education, and operates as a market-oriented international vocational training center.

At the initial stage of the project, 18 training rooms including mechanical processing technology, electrical technology and communication technology were built in three training centers for mechatronics and communication technology majors, with a total construction area of 6,814 m^2 and more than 1,600 sets of equipment.

The project will implement a "three-phase" plan. Based on the specialty construction in the first phase, international tourism, logistics engineering, automobile maintenance, building electricity and other specialties will be set up in the second phase to carry out technical skills training for Chinese&Cambodian enterprises and Cambodian people. Meanwhile, higher vocational education, applied technology undergraduate education, joint postgraduate education and other academic educations will be carried out to explore systematic talents cultivation of "medium and high vocational education, undergraduate education, and postgraduate education for a master's and doctoral degree".

Since 2017, as many as 95 articles about the project have been published by mainstream medium including People's Daily, Guangming Daily, China Education News, Xinhuanet, etc. from home and abroad. After over two months of field study and research, Tianjin Television produced two feature stories named "Khmer Training", each lasting 30 minutes. The two episodes were broadcast on May 6th and May 13th 2019 respectively, featuring "on and on sails the vocational education, overseas shines the Luban Workshop". They give a full coverage of how TSGUAS teachers brought advanced skills to local areas and how friendship flourished along "the Belt and Road Initiative", Which is a great contribution to "the Belt and Road Initiative". On July 18th, 2019, the Royal Government of Cambodia conferred the Officer of the SAHAMETREI Medal to the Secretary of the Party Committee of TSGUAS, and the Knight of the SAHAMETREI Medal to the President and Vice President in charge of this project, with the signature of Prime Minister Hun Sen of Cambodia. On July 22, 2019, China Education Association for International Exchange awarded TSGUAS the medal of "Featured Cooperation Project of China-ASEAN Higher Vocational Colleges". In October 2019, the President of National Polytechnic Institute of Cambodia (NPIC) presented 11 teachers with certificates and medals for their outstanding contributions to the Ministry of Labor and Vocational Training of Cambodia. In December 2020 Tianjin Sino-German University of Applied Sciences together with National Polytechnic Institute of Cambodia (NPIC) and their partners with enterprises was approved as the Belt and Road Joint Laboratory (Research Center), namely Tianjin Sino-German and Cambodia Intelligent Motion Device and Communication Technology Promotion Center .

The center has become a training base in Langcang-Mekong areas for technical talents training, a talent support base for Chinese enterprises overseas, a demonstration base for international students, and a base for teachers training. The Center is a key educational project of the Ministry of Foreign Affairs of the People's of China to serve "the Belt and Road Initiative" with foreign participation and entity institutions involved locally. The project will serve the social-economic and cultural development of the countries along the Initiative, enhancing the well-being of mankind. It will also serve the production output capacity of Chinese enterprises to help national development as well as enhance the international development of vocational education and the quality of its connotation. The project is a bridge connecting vocational education of Tianjin with the world, which marks a new stage of the city's international exchange and cooperation from a lower-medium to a medium-higher level.

The team of the project has compiled a series of textbooks based on current human resources situation in Langcang-Mekong countries, China's teaching equipment, and Chinese vocational qualification standards. These textbooks, involving six occupations like electrotechnics, lathe, milling, CNC operation, bench and 4G communication network, from elementary, intermediate to advanced level. These textbooks target competence development and orient students to work tasks, combining theory with practice, and learning with practicing so as to put knowledge and skills into real situations. The textbooks aim to provide skills standards for the six occupations and lay foundations for the upgrading of the technological level of Lancang-Mekong countries.

ZHANG Xinghui

Party Secretary of Tianjin Sino-German University of Applied Sciences

March, 2021

A Series of Textbooks for Lancang-Mekong Vocational Education Training Center and Cambodia Luban Workshop Editorial Committee

Preface

In order to better serve "the Belt and Road initiative" of China and Lancang-Mekong Vocational Education Training Center, and export high-quality technical skills & education concepts to the countries along Lancang-Mekong areas, teachers of the center have written this book based on the vocational skills standard and advanced equipment. This book based on the requirements of *National Vocational Skill Standard for Fitters* (*Elementary*) in terms of primary theoretical knowledge and related skills, combines the most advanced training equipment in the field of domestic vocational education, and flexibly adopts the systematic education theory of the work process. All the knowledge and skills are fully reflected in each module and specific work task. The book has made scientific innovations and breakthroughs in content and structure. Guided by practical work tasks and oriented by ability cultivation, the book is compiled according to the principle of project-based teaching mode, and combines theory with practice, learning and practice according to the actual task requirements, thus applying knowledge and skills into practice.

The modules and working content designed in this book extract from real cases and can meet all the regulations of primary technical knowledge and related skills requirements of *National Vocational Skill Standard for Fitters (Elementary)*. Written bilingually in both Chinese and English, the book is in line with international textbooks. It can not only for domestic readers who are seeking for professional English but also for overseas readers for their skills training or self-study.

Mi Jing of Tianjin Sino-German University of Applied Sciences acted as the chief editor; Song Quansheng, Ren Huifang and Tian Jiafeng of Tianjin Sino-German University of Applied Sciences acted as the deputy editor.

We greatly welcome the opinions and suggestions from readers and colleagues.

Editors

September 2023

Contents

Module One Basic knowledge of the fitter

Task One Safe awareness of the fitter

※ Knowledge objectives and skill objectives

(1) Understand the layout of the fitter's workplace.

(2) Master the knowledge of safe and civilized production of the fitter.

(3) Master the 5S management of workshop.

I. Workplace of the fitter

The workplace of the fitter refers to their fixed workplace. For the convenience of work, the layout of the fitter's workplace must be reasonable and meet the requirements of safe and civilized production.

(1) The main equipment should be arranged reasonably.

◇ The workbench should be placed in a place where there is proper light and it is convenient to work. The distance between workbenches should be appropriate. Workbenches which are placed face to face should be equipped with safe nets in the middle.

◇ The grinder and drilling machine should be installed on the edge of the workplace, and especially the grinder, it must be installed in a safe and reliable place.

(2) The workblanks and workpieces should be separated.

The workblanks and workpieces should be placed neatly, and the workpieces should be placed on the shelf as much as possible to avoid bumping.

(3) The machining tools, fixtures and measuring tools should be placed reasonably.

The commonly machining tools, fixtures, and measuring tools should be placed near the workplace for easy access at any time. After use, they should be maintained in time and put back to the original place.

(4) The workplace should be kept clean and tidy.

The equipment should be cleaned and lubricated as required after work, and the workplace should be cleaned.

II. General knowledge of safe and civilized production

Observing labor discipline, implementing safe operation procedures, and following the processing requirements strictly are important prerequisites for ensuring products quality. Safety cannot be separated from production. General knowledge of safe and civilized production includes:

(1) Wear protective equipment as required before work.

(2) Do not use unfamiliar machining tools and measuring tools without permission.

(3) The workblanks and semi-finished products should be placed in order according to the regulations, and oil stains and foreign matters should be removed as needed.

(4) Do not directly pull or wipe the cuttings chips by hand.

(5) Machining tools, fixtures and measuring tools should be placed at designated location, and it is strictly prohibited to place them randomly.

(6) Strictly observe the safe operation regulations of the fitter during work. The safe operation procedures for the fitter as shown in Table 1-1.

Table 1-1 Safe operation procedures for the fitter

Number	Content of safe operation procedures for the fitter
1	Keep dressed in work clothes, work shoes and work cap in the training room. Long hair must be tied into the cap. High heels, slippers, vests, skirts and shorts are forbidden to wear
2	Before work, check whether the workplace and tools are safe or not. There are any safety problems or damages, please clean and repair them in time and place them properly
3	The handle should be installed when using a file. The handle of the file must be free of cracks, with hoops around and with no iron wires
4	The file should not be placed outside the workbench, and it is not allowed to pry, smash or knock other objects
5	The workbench should be kept clean. To ensure the accuracy of the working surface of the workbench, and surface scratches should be avoided when moving
6	The saw blade should not be too loose or too tight, so as not to break. When the workpiece is clamped with a vise, the sawing position should not be extended too long
7	When using the tap and die, the workpiece must be processed vertically, with even force, not too strong, so as to prevent the damage of tool and workpiece. Be especially careful when threading the unpenetrated screw thread

Continued

Number	Content of safe operation procedures for the fitter
8	Do not reverse when threading, so as not to break the blade
9	The speed of the drilling machine cannot be changed arbitrarily. If it needs to be adjusted, get the teacher's approval first, stop the machine and then do the adjustment
10	When drilling, the workpiece must be clamped with the vise. It is strictly forbidden to hold the workpiece by hand. When the workpiece is about to be penetrated, be very careful and do not use too much force
11	The use of spanners and screwdrivers in the assembly should comply with the regulations. Do not use too much force to prevent accidents caused by skidding. The spanner used should meet the requirement of the nut. Stand in a proper position, and mind people nearby, prevent the spanner from skidding and hurting others. Do not use the spanner as a hammer

III. 5S management of workshop

5S refers to the five items of sorting (SEIRI), arranging (SEITON), sweeping(SEISO), cleaning (SEIKETSU), and quality (SHITSUKE). The first letter of each Japanese word is "S" in Roman letter, therefore it is called 5S for short. 5S originated in Japan. By standardizing the scene and the objects, it creates a clear working environment and cultivates good working habits of employees. Its ultimate goal is to equip people with the following qualities:

(1) Get rid of carelessness and develop the habit of taking everything seriously (take every small matter in the work seriously).

(2) Abide by the rules.

(3) Consciously maintain the good habits of clean and clear working environment.

(4) Keep civilized and polite habits.

5S includes:

(1) Sorting.

◇ Distinguish anything in the workplace as necessary and unnecessary.

◇ Clearly and strictly distinguish the necessary from the unnecessary.

◇ Dispose of unnecessary things as soon as possible.

(2) Arranging.

◇ Place and arrange in order the necessary articles left on the site after sorting.

◇ Define the quantity and mark them effectively.

(3) Sweeping.

◇ Sweep the workplace.

◇ Keep the workplace clean and bright.

(4) Cleaning.

◇ Institutionalize and standardize the implementation of above 3S(sorting, arranging and sweeping), implement and maintain the results.

(5) Quality.

◇Through morning meeting and other means, improve the civilization and politeness level of the whole staff. Cultivate each member to develop good habits and follow the rules. It is easy to carry out 5S, but long-term maintenance must rely on the improvement of quality.

Task Two Basic theoretical knowledge

※ Knowledge objectives and skill objectives

(1) Understand the definition and duties of the fitter, and master the classification of basic operation of the fitter.

(2) Master the basic knowledge of drawing recognition and be able to read the drawings of basic parts.

(3) Master knowledge of metal cutting.

(4) Master common knowledge of metal materials and heat treating.

I. Basic knowledge of the fitter

(I) Definition of the fitter

The fitter is a type of work that uses fitter tools or equipment to mark and process the workpieces, assemble and debug the machines, install and maintain the equipment, and manufacture and repair the tools according to the requirements. The fitter mainly works by hand, using various tools and common equipment to process metal. With the rapid development of industry, machine production has basically replaced manual labor in practical work, but in some occasions where machining is not suitable for the task, the fitter is still needed to complete the task, such as the production of complex and precise parts, assembly and maintenance of equipment, etc.

(II) Duties of the fitter

The main task of the fitter is to process, assemble and maintain the mechanical equipment. A machine is made up of many different parts. After being processed by various processing means, these parts need to be assembled by the fitter. In the assembly process, some parts often need to be assembled after supplementary operation by the fitter, such as drilling, tapping, key matching and pin processing. Some parts with low accuracy can achieve higher assembly accuracy after careful repair by the fitter. In addition, machines that hav been used for a long-time inevitably suffer from natural wear or damage from accidents which requires the fitter to repair. Therefore, the fitter's task is diverse and requires lots of professional knowledges.

With the development of machining and the improvement of production efficiency, the fitter's skills have become more and more complex, and its application range is becoming wider and wider. Due to the wide application of fitter skills, the fitter has a professional division of labor, such as assembling fitter, machine repairing fitter, tool fitter, etc., to meet the needs of different work and different occasions.

(III) Basic operations of the fitter

The basic operations of the fitter include: marking; filing; chiseling; sawing; drilling, counterboring, countersinking and reaming; tapping and threading; scraping and grinding; and assembly; etc.

1. Marking

Marking refers to using marking tools reasonably, and marking out the processing limit of the parts on the workpiece according to the marking procedure, which serves as the basis of the parts processing. As the first procedure of parts processing, marking has a close relationship with the processing quality of parts. The fitter should be familiar with the drawing, when marking out.

2. Filing

The filing processing goes as follows: use files of various shapes to file, reshape and repair the workpiece, so that the workpiece can reach a higher accuracy and a more accurate shape. Filing is one of the main operation methods of the fitter's work. All kinds of surfaces can be filed by this method.

3. Chiseling

Chiseling is to use simple tools such as the chisel and the hammer to cut the workpiece. Chiseling is the most basic operation of the fitter. It is used in situations where the requirements of parts processing are not high or the machining is impossible to work. Proficient chiseling skill is also an essential basic skill in fitter assembly and repair, which is widely applied.

4. Sawing

Sawing is used to cut the materials or cut the groove as required on the workpiece. When sawing, the saw blade and sawing method must be selected correctly according to the material property and the shape of the workpiece, so that the sawing operation can be carried out smoothly and meet the specified technical requirements.

5. Drilling, counterboring, countersinking and reaming

Drilling, counterboring, countersinking and reaming are four methods of rough machining, semi-finish machining and finish machining the hole. The drilling, counerboring and countersinking are carried out on the drilling machine. The reaming can be done by hand or by machine. Therefore, in order to master the operation skill of drilling, counterboring, countersinking and reaming, it is necessary to be familiar with the function of the cutting tools of drilling, counterboring, countersinking and reaming, as well as the structure and performance of the drilling machine and some machining tools and fixtures. In addition, in order to ensure the processing quality of drilling, counterboring, countersinking and reaming, it is necessary to master the specific methods of manual operation.

6. Tapping and threading

Tapping and threading are to process the internal thread or external thread on the inner hole or outer cylinder surface of the workpiece with a tap and a circular die. Tapping and threading are usually used by the fitter. The screw thread processed by the fitter is usually the thread with a smaller diameter or unsuitable for machining on the machine tool. In order to make the processed thread meet the technical requirements, the fitter not only should be familiar with the formation of the thread, the size relationship of each part, and the tool for cutting the thread, but also master the key points of thread processing operation and the methods to avoid waste products.

7. Scraping and grinding

Scraping is a method used by the fitter to finish the workpiece. After scraping, the surface of the workpiece can not only reach the geometric accuracy, dimensional accuracy, contact accuracy and transmission accuracy, but also make the surface structure closer through the scraper's squeezing the surface of the workpiece during the scraping process, so as to improve the mechanical properties.

Grinding is the most accurate processing method. When grinding, the abrasive can slide and roll between the lapping tool and the workpiece to produce micro cutting, that is the physical role in grinding. At the same time, the chemical role of some abrasives is used to produce an oxide film on the surface of the workpiece, but the oxide film itself is easy to be ground off during grinding. In this way, the oxide film is constantly produced and abraded, so that the surface of the workpiece have a high accuracy.

8. Assembly

Assembly is the process of combining parts into bigger units or complete machines through appropriate connection forms according to the technical requirements specified in the drawings.

(IV) Characteristics of the fitter's work

The characteristics of the fitter's work can be summarized as follows:

(1) It is flexible and convenient to process, and can able to process parts which have complex shapes and have high quality requirements.

(2) The tools are simple, the manufacturing edge is easy to sharpen, the material source is sufficient, and the cost is low.

(3) The fitter have high labor intensity, low production efficiency and high requirements for individual skill levels.

II. Knowledge of drawing recognition

In modern industrial production, machinery, chemical industry or buildings are manufactured and constructed according to drawings. Designers express design intention through drawings; manufacturers understand design requirements, organize manufacturing and guide production according to drawings; users understand the structure and performance of machinery and equipment through drawings, and carry out operation, repair and maintenance. Therefore, mechanical drawing is a medium and tool to exchange and transmit technical information and ideas, and it is a common technical language in the engineering field. The modern skilled talents at the first line of production which comply with the target of vocational and technical education, must learn and master this language, and have the basic ability to read and draw mechanical drawings.

(I) Principle of orthographic projection

Assuming the light source is located at an infinite distance, the projection of its projection lines on the projection plane after passing through an object is called orthographic projection. Its projection lines are parallel to each other and perpendicular to the projection plane.

(II) Three-view drawing

An object in a certain space cannot be completely and accurately expressed by the one-view projection, therefore, three-plane projection system which is composed of three mutually perpendicular projection planes should be established. In the system, the *V* plane is the vertical plane, which is briefly called the front; the *H* plane is the horizontal plane, which is briefly called the horizontal plane; the *W* plane is the side vertical plane, which is briefly called the side (see Fig. 1-1). The projection of geometric elements in the *V*, *H* and *W* three-view projection system is called the three-plane projection of geometric elements.

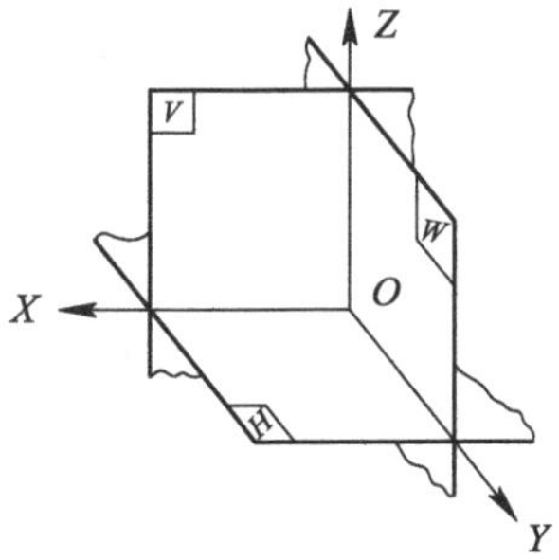

Fig. 1-1 Three-plane projection system

Put an object in the three-plane projection system, project it to three projection planes respectively, unfold the three projection planes to get one plane, and then three views of the object can be obtained.

When drawing a mechanical drawing, the figure obtained by orthographically projecting the object onto the projection plane is called a view. The front projection in the three-view projection system is called the front view; the horizontal projection is called the top view; the side projection is called the left view. This three views are collectively called the three-view projection of the object as seen in Fig. 1-2.

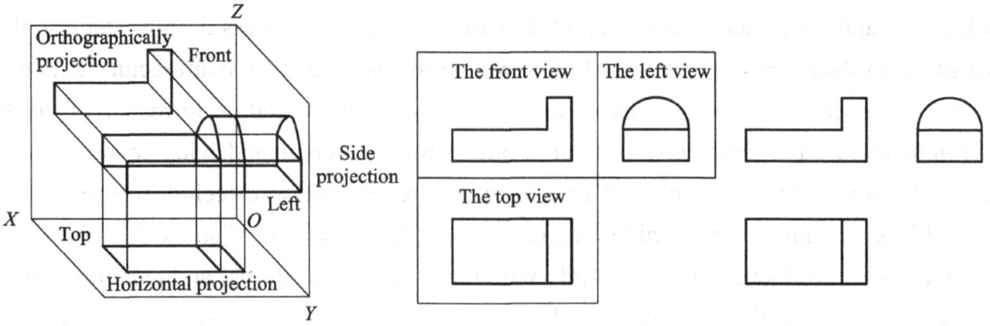

(a) Three-view projection (b) Three-view drawing (c) Three-view drawing of the actual picture

Fig. 1-2 The formation of three-view drawing

(III) Methods of reading the three-view drawing of the basic geometric solid

The position relationship of the three-view drawing is as follows: the top view is below the front view, and the left view is to the right of the front view.

The front view reflects the position relationship between of the object top to down and left to right, that is the height and the length of the object.

The top view reflects the position relationship between of the object left to right and front to back, that is the length and the width of the object.

The left view reflects the position relationship between of the object top to down and front to back, that is the height and the width of the object.

Therefore, the projection rules of the three-view drawing can be summarized as follows: the length of the front view equals that of the top view, the height of the front view equals that of the left view, the width of the top view equals that of the left view, as shown in Fig. 1-3. The projection rules of the three-view drawing is abbreriated as three-equrallence rule.

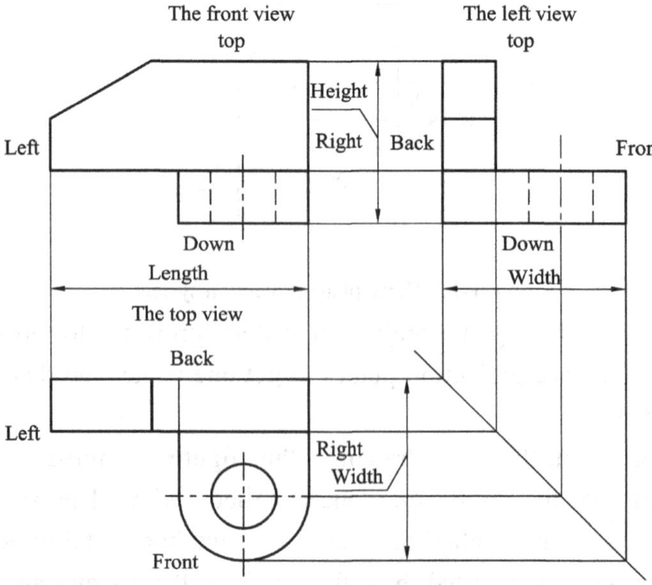

Fig. 1-3 The position and projection relationship of three-view drawing

1. Basic methods of reading the drawing

Reading the drawing is still mainly based on the shape analysis and supplemented by the line-surface analysis. According to the view of the shape, each shape is identified one by one, and then the combination of the shapes and the relative position of the adjacent surfaces between the shapes are determined. After initially constructing in mind the compound solid, it is also necessary to verify whether each given view matches the view of the imaginary compound solid. When they are inconsistent, the imaginary view must be corrected according to the given view until each view is consistent. Then, the imaginary compound solid is the desired one.

2. Key points of reading the drawing

(1) Start with the views that reflect the features of the shape, and connect several views together.

It takes two or more views for a compound solid to be expressed clearly. Among the views, the front view is the one that best reflects the shape characteristics of the compound solid and the relative position of each other. Therefore, when reading the drawing, start from the front view, and connect several views together. Only by this, the shape and the relative position between the solids can be accurately identified. Never jump to a conclusion after reading one view. As can be seen from Fig. 1-4, the shape of the compound solid cannot be determined by only one view.

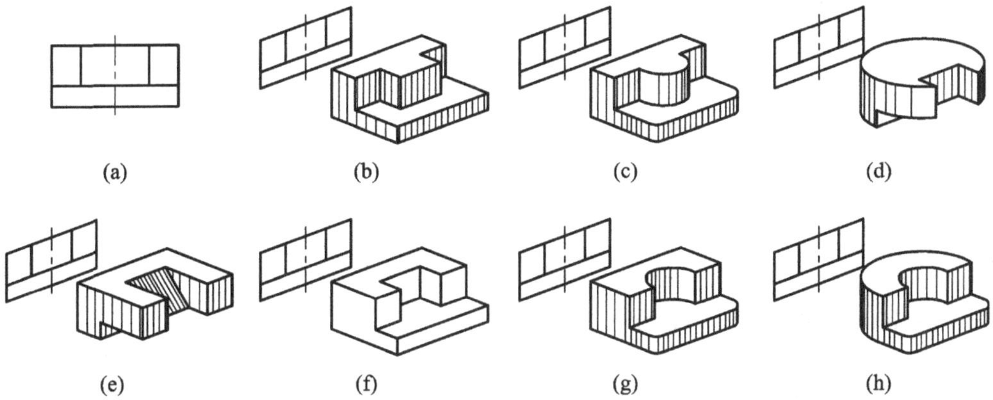

Fig. 1-4 The shape of the compound solid cannot be determined by only one view

(2) The frame in the view should be carefully analyzed to identify the position of the shape and the surface of the shape.

As shown in Fig. 1-5, when several frames are connected or there are wire frames in the wire frame of a certain view of the compound solid, by comparing the projection relationship, the position relationship of the front and back, top and down, left and right and intersection can be distinguished, which help to imagine the shape.

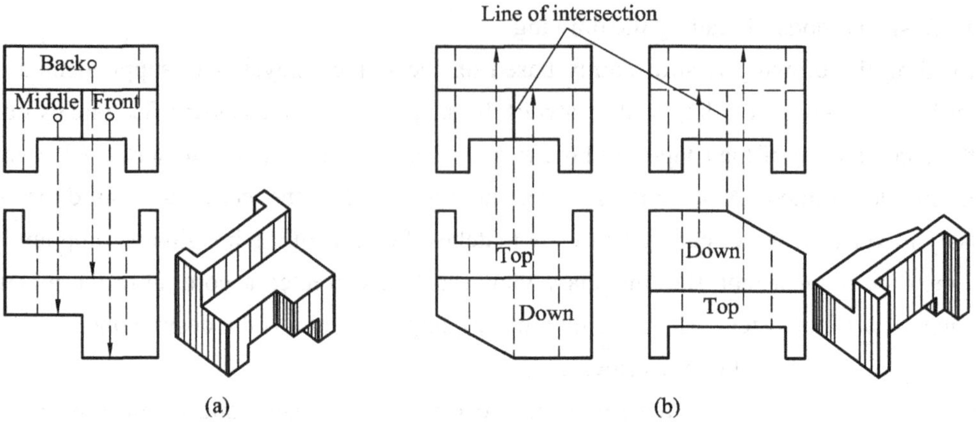

Fig. 1-5 Determining the position of each surface

(3) Compare the imaginary shape with the given view repeatedly.

Reading the drawing is the process of constantly comparing the imaginary compound solid with the given view. In other words, reading the drawing is the thinking process of constantly modifying the compound solid in imagination.

3. Steps of reading the drawing

(1) Divide the wire frame and compare projection relationships. With the help of T-square, set square and compass, and according to the projection principles of the three-view drawing, a comprehensive analysis of several viws is carried out from the front. The compound solid is roughly divided into several parts.

(2) Identify the shape and position. Imagine the shape from the view of each part and determine their position.

(3) Think of them as a whole. After determining each shape and the position, the shape of the whole compound solid is revealed. Then it's better to check one by one the compound solid imagined in the process with the given three-view drawing.

III. Knowledge of metal cutting

Metal cutting is a kind of machining method that uses the cutter to cut off the surplus metal materials on the workpiece to obtain parts that meet the requirements, such as turning, filing, drilling, planing, cylindrical grinding, surface grinding, etc.

(I) Cutting movement

Cutting movement is refer to the relative movement between the tool and the workpiece It is divided into two types: main movement and feed movement.

1. Main movement

The main movement is the most basic movement for cutting, which is refers to the workpiece and the cutter move relatively to cut. There is only one main movement, which consumes the most power. It can be rotary movement, such as the rotation of workpiece in turning, the rotation of drill bit during drilling, and the rotation of grinding wheel during grinding.

2. Feed movement

The feed movement refers to the movement of gradually cutting out the entire surface of the workpeice in order to maintain the continuous cutting. There are one or more feed movement, which consumes less power, and can be continuous or intermittent.

3. Workpiece surface during cutting

(1) Surface to be machined: the surface to be cut.

(2) Machined surface: the surface of a workpiece that is formed after being cut by a tool.

(3) Transition surface: the surface of a workpiece that is being cut by a cutter.

(II) Cutting parameters

Cutting parameters is a collective term of cutting speed, feed rate and cutting depth in the cutting process. Generally, they are called "three elements of cutting".

1. Cutting speed (v_c)

The cutting speed refers to the instantaneous speed (m/min) of the selected point on the cutting edge of the tool relative to the surface to be machined (i.e. the linear speed of the movement). The cutting speed is selected according to the size accuracy of workpiece, surface roughness and tool service life, and is determined by calculation, table checking or experience.

2. Feed rate (f)

The feed rate refers to the relative displacement (mm/r) between the workpiece and the cutter along the feed motion direction in a working cycle of the main movement (The workpiece or the cutter rotates once or reciprocates once). The feed rate has a great influence on the surface roughness. f should be larger in rough machining and smaller in finishing machining.

3. Cutting depth (a_p)

Cutting depth refers to the vertical distance (mm) of the machined surface on the workpiece. During rough machining, after leaving the finishing machining allowance, the remaining machining allowance should be used up as soon as possible. If the allowance is too large, it can be cut several times. For the first time, the a_p value should be as large as possible. When finishing, the appropriate a_p value should be selected according to the machining accuracy and surface roughness.

The basic principles of selecting cutting parameters is as following: under the premise that the rigidity of machine tool, workpiece, tool and process system is allowed, the first choice is the large cutting depth, the second is the large feed rate, and the last is the large cutting speed.

(III) Metal cutting tool

1. Cutting tool composition

The cutting tool is generally composed of cutting part (tool bit) and clamping part (cutter body). The cutting part consists of three parts: three sides, two edges and tool bit. Three sides refers to the front (rake face) and main back (flank face); two edges refers to the main cutting edge and auxiliary cutting edge; tool bit refers to the intersection point of the main cutting edge and the auxiliary cutting edge.

2. Cutting tool material

The material of the cutting tool should have the following properties.

(1) High hardness. The hardness of the cutting part must be higher than that of the workpiece material.

(2) High wear resistance. Generally, the higher the hardness, the better the wear resistance.

(3) Sufficient strength and toughness. Strength and toughness refers to the ability to withstand pressure, impact and vibration in the process of use, so as to prevent the tool from chipping or brittle fracture.

(4) High heat resistance. High heat resistance means that the tool material can maintain high hardness at high temperature, also known as red hardness or thermal hardness.

(5) Good processability. In order to facilitate the tool processing, the cutting tool material should have good machinability and heat treatment.

3. Cutting tool materials commonly used by the fitter

(1) Carbon tool steel, such as T10A, T12A, can be used to manufacture low-speed hand tools, such as files, chisels and saw blades.

(2) High speed steel wear resistance and heat resistance are significantly improved than T10A and T12A. They can withstand the temperature of 550–600℃, such as turning tool, milling cutter, drill bit, reamer, broach, gear cutter and finishing tool, etc.

(3) Cemented carbide has high wear resistance and heat resistance. It's hardness is still very high at the temperature of 800–1000°C. The disadvantage is that it is inferior to high-speed steel in strength, toughness and processability.

(4) Alloy tool steel, such as 9SiCr and CrWMn, has high toughness, wear resistance and heat resistance, and can be used to make taps and dies.

4. Tool wear and tool life

1) Forms of tool wear

(1) Rake face wear: It refers to the crescent groove on the rake face when cutting plastic materials or when the cutting speed and cutting depth are large.

(2) Flank face wear: When machining brittle materials or machining at low cutting speed and with small cutting depth ($<$0.1 mm), the wear occurs on the flank face due to the relatively weak interaction between the chips on the rake face.

(3) Boundary wear: When cutting steel, the deep grooves are often formed at the main cutting edge near the workpiece skin and on the flank face of the tool bit.

2) Causes of tool wear

(1) Abrasive wear: It refers to the phenomenon of surface material loss caused by the friction between the object surface and hard particles or hard protrusions (including hard metals), which is the main wear reason in low-speed cutting.

(2) Adhesive wear: It refers to the surface damage scar caused by cold welding bonding between chip material and tool under pressure and friction conditions and taking away the tool

material. Under the condition of unstable accretion, the adhesive wear is very serious.

(3) Diffusion wear: It refers to the tool wear caused by the mutual diffusion of workpiece and tool materials at high temperature. One of the main causes of carbide tool wear is diffusion wear.

(4) Chemical wear: It refers to the wear caused by the chemical action between the tool material and the surrounding media at high temperature, and a layer of low hardness compound formed on the tool surface is wiped off by the chip or workpiece. Chemical wear is the main reason of boundary wear in high speed cutting.

3) Tool life

Tool life refers to the cutting time from the beginning of cutting to the time when the wear reaches the blunt standard (VB, NB). It is expressed in T and the unit is minute. Tool life is an important parameter to measure the cutting performance of tool material, the machinability of workpiece material and the rationality of tool geometric parameters. Among the cutting parameters, the cutting speed has the greatest influence on the tool life, followed by the feed rate and the back engagement of the cutting edge.

(IV) Cutting fluid

1. The function of cutting fluid

(1) Cooling effect: reduce cutting temperature, prolong tool life and improve production efficiency.

(2) Lubrication: form lubricating film, reduce friction and improve surface quality.

(3) Cleaning function: wash the chips and abrasive particles, so as not to affect the surface quality of the workpiece and the accuracy of the machine tool.

(4) Anti rust effect: add anti-rust agent to protect the workpiece, the cutter and the machine tool.

2. Classification of cutting fluid

(1) Water soluble cutting fluid: mainly used for cooling.

(2) Oil soluble cutting fluid: mainly used for lubrication.

IV. Knowledge of common metal materials and heat treatment

(I) Properties of metal materials

Generally speaking, the properties of metal materials include physical properties, chemical properties, mechanical properties and process properties.

1. Physical and chemical properties of metal materials

(1) Density. The mass per unit volume of a substance is density. Density is represented by the symbol ρ. The concept of material density can be used to solve a series of practical problems, such as calculating the weight of workblank and identifying metal materials. The density of common metal materials is as follows: the density of cast steel is 7.8 g/ cm^3, the density of gray cast iron is 7.2 g/ cm^3, the density of steel is 8.9 g/ cm^3, the density of brass is 8.63 g/ cm^3, the density of aluminum is 2.7 g/ cm^3.

(2) Electrical conductivity. The metal's ability to conduct an electric current is electrical conductivity. Various metals are different in electrical conductivity. Silver usually has the best conductivity, followed by copper and aluminum.

(3) Thermal conductivity. The metal's property of conducting heat is thermal conductivity. Generally speaking, materials with good electrical conductivity have good thermal conductivity. If some parts need a lot of heat absorption or heat dissipation in use, materials with good thermal conductivity should be used. For example, the cooling water pipe in condenser is usually made of copper alloy with good thermal conductivity to improve the cooling effect.

(4) Thermal expansion. When the metal is heated, its volume expands. This phenomenon is the thermal expansion of metal. For example, the uneven thermal expansion of the workpiece due to uneven heat will lead to welding deformation and welding stress. The index of thermal expansion is coefficient of thermal expansion.

(5) Oxidation resistance. The metal's ability to resist the corrosive action of oxidizing gases at high temperature is oxidation resistance. High temperature parts in thermal equipment, such as the boiler superheater, the water wall tube, the cylinder and the blade of steam turbine, are easy to produce oxidation corrosion. Generally, the anti-oxidation corrosion rate index of the material used as superheater tube should be controlled at $\leqslant 0.1$ mm / a.

(6) Corrosion resistance. The metal's ability to resist corrosion of various media (atmosphere, acid, alkali, salt, etc.) is corrosion resistance. Many parts of chemical and thermal equipment work for a long time under corrosive conditions, so the corrosion resistance of steel must be considered when selecting materials.

2. Mechanical properties of metal materials

When metal materials are subjected to external load, the mechanical characteristics during the whole process from the beginning of stress to material failure are called mechanical properties. Mechanical properties is an important index to measure the performance of metal materials. The mechanical properties mainly include strength, plasticity, hardness and impact toughness.

1) Strength

The strength performance of metal material represents the resistance of metal material to deformation and fracture. It is expressed by the force on unit section (called stress). The common strength indexes are yield strength and tensile strength.

(1) Yield strength. In the tensile process, when the tensile stress reaches a certain value and no longer increases, its deformation continues to increase. This tensile stress value is yield strength, expressed by σ_s. The higher the value of σ_s, the higher the strength of the material.

(2) Tensile strength. The maximum tensile stress of a metal material before failure is tensile strength. It expressed as σ_b. The greater the value of σ_b, the greater the fracture resistance and the higher the strength.

2) Plasticity

Plasticity refers to the metal's ability to produce plastic deformation under the action of external forces. The plastic properties of metal materials include elongation, reduction of area and cold bending angle.

(1) Elongation. Elongation is refers to the percentage of elongation and original length when a metal material is broken by tensile force. It is expressed by δ. Elongation $\delta=[(L_1-L_0)/L_0]\times 100\%$. In the equation, L_0 refers to the original length of the specimen and L_1 is the length of the specimen after being broken.

(2) Area reduction. Area reduction is refors to the percentage of the reduced cross-sectional area at the breaking point and the original cross-sectional area when a metal material is broken by tensile force. Area reduction $\varphi=[(A_1-A_0)/A_0]\times 100\%$. Here A_0 stands for the original cross-sectional area of the specimen and A_1 is the cross-sectional area of the necking part after being broken.

The area reduction is not affected by the gauge length of the specimen, so it can reflect the plasticity of the material more reliably. For materials that must bear strong deformation, the material with good plasticity has better performance after cold pressing.

(3) Cold bending angle. Cold bending angle is also called bending angle. Generally, long strip specimens are bent by the specified bending radius according to different materials and plate thickness. When cracks appear on the tensile surface, the angle between the specimen and the original plane is cold bending angle. It is represented by α. The larger the cold bending angle is, the better the plasticity of the metal material is.

3) Impact toughness

Impact toughness is the metal's ability to resist dynamic load or impact force. Impact test can determine the sensitivity of materials to notch under sudden load. Impact value is an index of impact toughness, expressed by a_k. The larger the a_k, the better the toughness of the material. $a_k=A_k/S_n$, which is widely used in impact test of pressure bearing special equipment materials. Here A_k is the impact absorption energy, with the unit of J; S_n is the cross-sectional area of the sample groove before the test, with the unit of cm^2; a_k is the impact value, with the unit of in J/cm^2.

4) Hardness

Hardness is the metal's ability to resist surface deformation. The commonly used hardness indexes are Brinell hardness HB, Rockwell hardness HR and Vickers hardness HV.

3. Technological properties of metal materials

The technological performance of metal materials refers to the ability to withstand various cold and heat processing.

(1) Cutting performance. Cutting performance refers to whether metal materials are easy to cut. When cutting, if the cutting tool is not easy to wear, the cutting force is small and the surface quality of the workpiece to be cut is high, then the cutting performance of the material is considered to be good. Generally, gray cast iron has good cutting performance, and the hardness of steel is in the range of HB180–200 has good cutting performance.

(2) Castability. The metal's castability refers to the fluidity of metal in liquid state and the area reduction and segregation degree of liquid metal in solidification process. The metal's

castability is an important property to ensure the quality of casting.

(3) Weldability. weldability refers to the metal's ability to be welded into components according to specified design requirements under limited construction conditions and meet the predetermined service requirements. Weldability is affected by material, welding method, component type and service requirements.

There are many methods to evaluate weldability, among which carbon equivalent method is widely used. This method is based on the effect of alloy elements on the weldability of steel. The content of alloy elements (including carbon) in steel is converted into the equivalent content of carbon according to its function, which can be used as a reference index for evaluating the weldability of steel. The carbon equivalent method is used to estimate the hardening and cold cracking tendency of carbon steel and low alloy steel.

(II) Classification of steel

1. Classification by chemical composition

(1) Carbon steel: In addition to iron, this kind of steel mainly contains carbon, silicon, manganese, sulfur, phosphorus and other elements, the total amount of these elements is generally less than 2%.

According to the carbon content, carbon steel can be divided into the following categories:

◇ Low carbon steel: with carbon content less than 0.25%.

◇ Medium carbon steel: with carbon content of 0.25%–0.60%.

◇ High carbon steel: with carbon content more than 0.60%.

(2) Alloy steel: In addition to the elements contained in carbon steel, there are other elements such as chromium, nickel, titanium, molybdenum, tungsten, vanadium, boron, etc. If the content of manganese in the steel is more than 0.5%, or the content of silicon is more than 0.5%, it is also called alloy steel.

2. Classification by quality grade

(1) Ordinary steel: The sulfur content is not more than 0.045%–0.050%, and the phosphorus content is not more than 0.045%.

(2) High quality steel: The sulfur content is not more than 0.030%–0.035%, and the phosphorus content is not more than 0.035%.

(3) High quality steel: The sulfur content is not more than 0.020%–0.030%, and the phosphorus content is not more than 0.025%–0.030%.

3. Classification by usage characteristics

(1) Structural steel.

(2) Tool steel.

(3) Special purpose steel, such as stainless steel, acid resistant steel, heat resistant steel, magnetic steel, etc.

(III) Heat treatment of steel

Heat treatment of steel is a very important process in the manufacturing process of mechanical parts and tools. Heat treatment is to heat steel to a certain temperature, keep its temperature steady for a certain time, and then cool at a certain speed to change its internal structure (generally the chemical composition and shape don't change), so as to change the performance of steel. The operation process is divided into three basic stages: heating, heat preservation and cooling.

The main purpose of heat treatment is to:

(1) Improve hardness, strength and wear resistance.

(2) Reduce the hardness and improve the plasticity, so as to facilitate cutting.

(3) Eliminate the internal stress caused by steel processing.

(4) Improve the internal structure and properties of steel parts to meet the process requirements.

(5) Improve the wear resistance, corrosion resistance and oxidation resistance of metal surface.

According to the change of micro-structure and properties of steel during heating and cooling, the heat treatment process is divided into annealing, normalizing, quenching, tempering, chemical heat treatment, etc.

Task Three Common equipment and measuring tools for the fitter

※ Knowledge objectives and skill objectives

(1) Understand the types, specifications and uses of tools and equipment for the fitter.

(2) Understand the precautions of common tools.

(3) Understand the composition and structure of common measuring tools for the fitter.

(4) Be able to use the vernier caliper and the micrometer correctly.

I. Common equipment

Fitter work place is a fixed place for a group of workers to work. In this fixed place, the main equipment installed includes workbench, bench vice, grinding machine, bench drilling machine, etc.

1. The fitter's workbench

As shown in Fig. 1-6, the fitler's workbench is also named a fitter's bench or bench work table or bench work platform, whose main function is to install bench vices and store common tools used for machining, clamping and measuring.

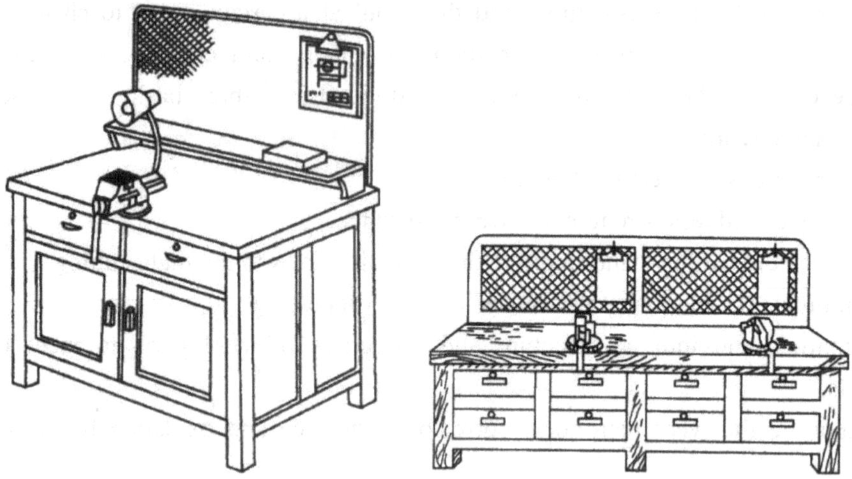

Fig. 1-6 The fitter's workbench

2. Bench vice

Bench vice is a general fixture used to clamp workpiece. Its specifications are expressed by the width of the jaws, and the common specifications include 100 mm, 125 mm and 150 mm.

There are two types of bench vice, fixed and rotary, as shown in Fig. 1-7. The main structure and working principle of them are basically the same. The difference is that the rotary bench vice has one more base than the fixed bench vice, and the body of the vice can rotate on the base during work. Therefore, it is convenient to use and has a wide range of applications, which can meet the processing needs of different directions.

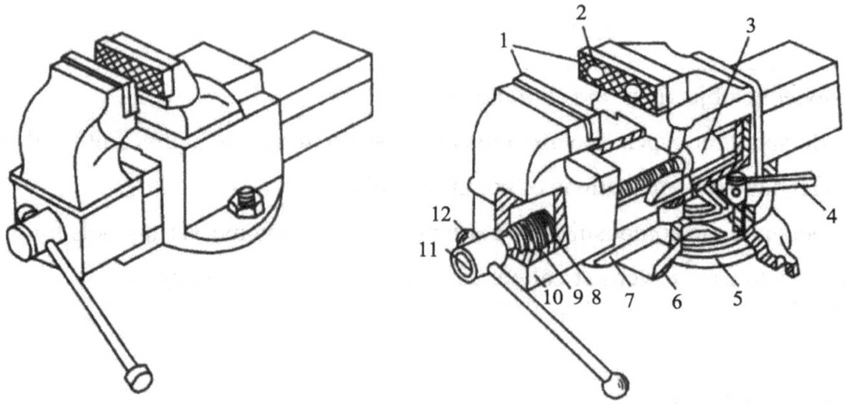

(a) The fixed type (b) The rotary type

1-Jaw; 2-Screw; 3-Nut; 4, 12-Handle; 5-Clamping plate; 6-Turntable base; 7-Fixed body; 8-Retaining ring; 9-Spring; 10-Movable body; ll-Lead screw.

Fig. 1-7 Bench vice

Attention should be paid to the following when using the bench vice:

(1) When clamping the workpiece, it must be tightened properly. The handle can only be tightened by hand, and do not use other tools to exert force.

(2) When working with force, try to make the force directed towards the fixed body.

(3) It is not allowed to knock on the movable body and smooth surface.

(4) Clean and lubricate the moving surface of the lead screw and nut frequently to prevent rust.

3. Grinding machine

Grinding machine is a common equipment used to sharpen various knives and other tools, which is composed of a motor, a grinding machine base, a bracket and a protective cover, as shown in Fig. 1-8.

Fig. 1-8 Grinding machine

The grinding wheel is brittle and its rotation speed is very high. The following safety operation rules should be strictly observed during work:

(1) The rotating direction of the grinding machine should be correct, so that the grinding debris can only fly downward from the grinding wheel.

(2) After the grinding machine is started, grinding should be carried out after the grinding wheel rotates smoothly. If the grinding wheel skips obviously, stop the machine for readjusting in time.

(3) The distance between the grinding machine bracket and the grinding wheel should be kept within 3mm to prevent accidents caused by workpiece rolling in. Before starting the grinding machine, carefully check whether there is any debris or other things attached or blocked between the grinding wheel and the protective cover. If there is, the debris or other things must be cleaned before starting the machine.

(4) When installing the grinding wheel, the clearance between the inner hole of the grinding wheel and the spindle should not be too close. It should be controlled within 0.1 mm according to the clearance requirements.

(5) When tightening the nut, a special wrench should be used and it cannot be too tight. It is

strictly forbidden to hammer with hard objects to prevent the grinding wheel from being broken.

(6) Don't rush to start the newly installed grinding wheel. First check it by inching, and then idle for 2 to 3 minutes. It can only be used after the grinding wheel runs normally.

(7) It is not allowed for two people to grind on the same grinding wheel at the same time. After the grinding operation, the power supply of the grinding machine should be turned off in time, and the grinding machine should not be left idling.

(8) Those who operate the grinding machine should wear safety goggles.

(9) The dust suction device of grinding machine must be intact and effective. If there is any problem, stop working and repair it in time.

4. Bench drilling machine

Bench drilling machine is called bench drill for short. It is simple in structure and easy to operate. It is often used to drill small workpieces and expand holes with diameter less than 12cm. The main structure of the bench drilling machine is shown in Fig. 1-9.

Fig. 1-9 Bench drilling machine

The safety operating procedures for bench drilling machines are as follows:

(1) Perform a comprehensive check of the drilling machine, clamping and measuring tools before working. It can only be started after being confirmed to be in good condition.

(2) It is strictly forbidden to wear gloves to operate machines, and girls should pull their hair braids in their hats.

(3) The installation and clamping of the workpiece must be firm and reliable. When drilling small parts, they should be held with a tool. It's forbidden to hold them by hand when drilling.

(4) When using automatic feed, it is necessary to select the proper feed speed and appropriately adjust the stroke limit block. During manual feed, generally follow the principle of gradual pressure increase and gradual pressure reduction to avoid accidents caused by excessive force.

(5) When there are long iron filings around the drill bit, stop and remove them. It is

forbidden to blow with mouth or pull with hand. It should be removed with a brush or an iron hook.

II. Operation of common equipments for the fitter

1. Operation and maintenance practice of bench vice

First of all, get to know the structure of the bench vice, be familiar with the function of each handle, and then perform basic action practice of clamping and loosening the workpiece, rotating and fixing the turntable, and the daily maintenance practice of the bench vice.

2. Operation of the grinding machine and grinding practice

Carefully observe the structure of the grinding machine, adjust the bracket to make sure that the distance between the bracket and the grinding wheel is not more than 3 mm, then practice grinding, and then carry out the practice of replacing the grinding wheel and the routine maintenance of the grinding machine.

3. Operation practice of the bench drilling machine

(1) Carefully observe the structure of the bench drilling machine, be familiar with the function of each handle, and practice lubrication.

(2) The main shaft speeds gradually from low speed to high speed for practice.

(3) Practice manual feed and master feed at a constant speed.

(4) Practice the lifting, lowering and fixing of the workbench.

(5) After mastering each single operation, perform practice of drill bit clamping, idling, feed and the maintenance of the drilling machine.

III. Common measuring tools for the fitter and their usage methods

In order to ensure the quality of products, it is necessary to strictly measure the workpiece during and after processing. The tools used to measure the shape and size of workpieces and products are called measuring tools or measuring instruments. There are many types of measuring tools. According to their uses and characteristics, they can be divided into many types such as universal measuring tools, special measuring tools and standard measuring tools.

(1) Universal measuring tools: They can be used to measure parts of various shapes and sizes. This kind of measuring tools generally have scales, which can measure the specific values of the shape and size of parts or products within the measuring range, such as vernier caliper, micrometer, dial indicator, etc.

(2) Special measuring tools: They are specially designed to measure parts or products of some special shape or size. This kind of measuring tools cannot measure the actual size, and they can only check if the shape and size of parts or products is qualified, such as caliper.

(3) Standard measuring tools: They can only be made into a certain fixed size for proofreading and adjusting other measuring tools, such as measuring gauge.

1. Vernier caliper

Vernier caliper can be used to measure length, thickness, outer diameter, inner diameter, hole depth and center distance. There are three kinds of accuracy of vernier calipers: 0.1 mm, 0.05 mm and 0.02 mm.

The structure of vernier caliper is shown in Fig. 1-10. It is composed of the main scale, sliding vernier scale, inner measuring jaws, outer measuring jaws, depth ruler and lock screw.

Firstly, read out the whole number in millimeter on the left side of main scale's zero line, and then see how many lines of the vernier caliper starting from the left zero line are aligned with a certain line on the ruler body. The number of the lines multiplied by accuracy stands for the decimal part which is less than 1 mm. Finally, add the whole millimeter number and decimal number, and the actual size can be obtained, as shown in Fig. 1-11.

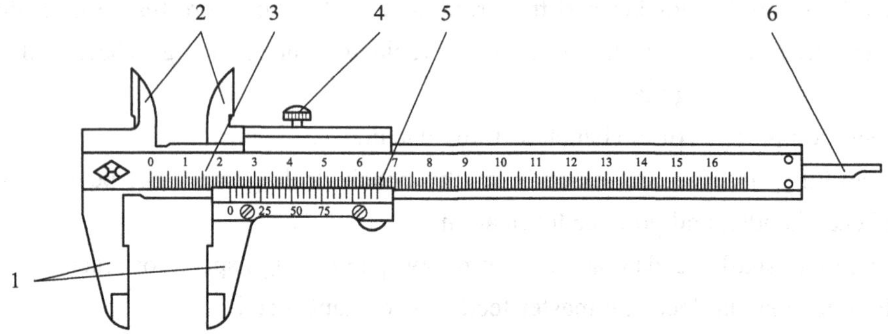

1–Outer measuring jaws; 2–Inner measuring jaws; 3–Main scale;
4–Lock screw; 5–Sliding vernier scale; 6–Depth ruler.

Fig. 1-10 Vernier caliper

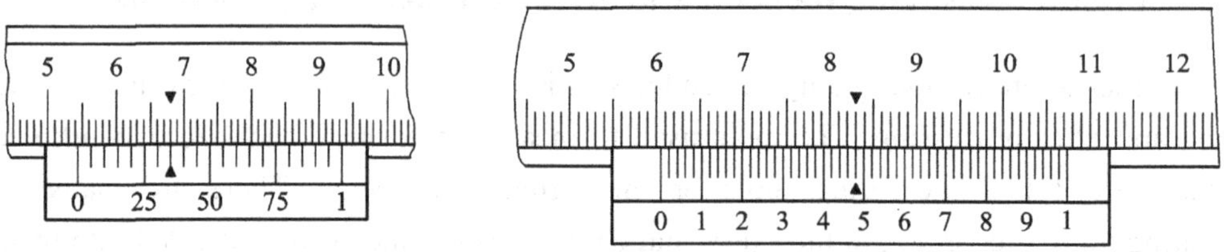

54 mm+0.35 mm=54.35 mm

(a) Reading method of 0.05 mm vernier caliper

60 mm+0.48 mm=60.48 mm

(b) Reading method of 0.02 mm vernier caliper

Fig. 1-11 Reading methods of vernier caliper

The precautions for using vernier calipers are as follows:

(1) Select an appropriate vernier caliper according to the size and dimensional accuracy of the workpiece. The vernier caliper is only suitable to measure and inspect the workpiece of the medium tolerance grade (IT10–IT16). The vernier caliper cannot be used to measure the size of the casting and forging workblanks, otherwise it will wear out quickly and lose its accuracy.

(2) Before use, the vernier caliper shall be checked to make sure that the measuring jaws are clean, the measuring faces and the measuring edges of the jaws are straight and undamaged, there shall be no light leakage when the two measuring jaws are closely jointed, and the zero line

of the main scale and the sliding vernier scale should be aligned.

(3) When measuring, the inner measuring jaws and outer measuring jaws should be opened to the extent that is slightly larger than the measured size. Firstly, stick the ruler frame to the measuring datum plane of the workpiece, and then gently move the vernier to make the inner and outer measuring jaws stick to the other side of the workpiece, and make sure that the measuring faces of the vernier caliper and the workpiece contact correctly, as shown in Fig. 1-12. It is not right for them to be placed in the position as shown in Fig. 1-13. Finally tighten the lock screw and read the value.

(4) When reading, keep the vernier caliper in a horizontal position and try to make sure the line of the person's sight be vertical to the scribed surface of the vernier caliper to avoid the reading error caused by the deviation of line of sight.

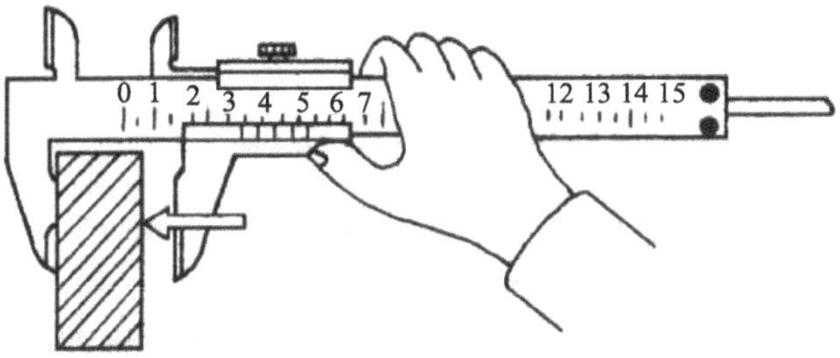

Fig. 1-12 The method of using the vernier caliper

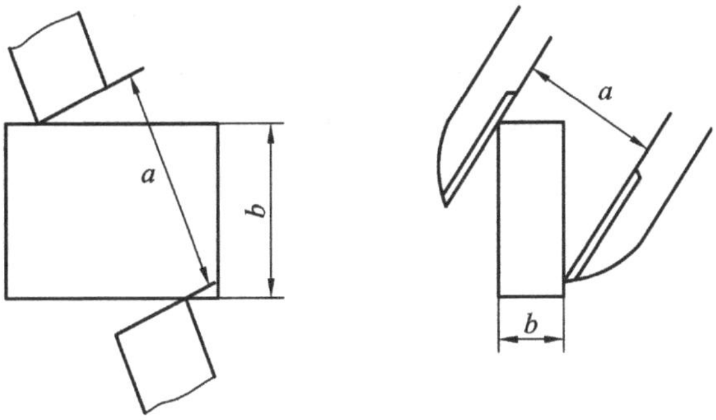

Fig. 1-13 Wrong contact between the measuring surface of vernier caliper and workpiece

2. Micrometer

A micrometer is a kind of accuracy measuring tool. It has higher accuracy than the vernier caliper and it is more sensitive. Therefore, the parts that require high machining accuracy, should be measured with a micrometer.

Micrometer can be divided into three kinds, namely, outside micrometer, inside micrometer and depth micrometer, based on their use and structure. Among them, the outside micrometer is the most commonly used. When the measuring range is within 500 mm, the specification is divided every 25 mm, such as 0–25 mm, 25–50 mm, etc.; When the measuring range is 500–1000 mm, the specification is divided every 100 mm, such as 500–600 mm, 600–700 mm, etc.

1) Configuration of the micrometer.

Fig. 1-14 shows the configuration of the micrometer.

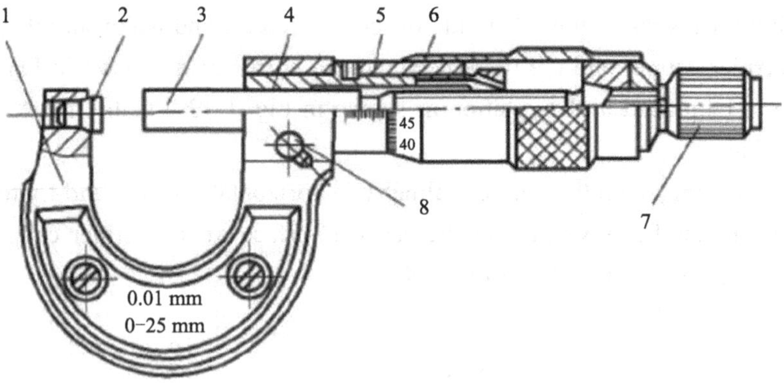

1–Frame; 2–Anvil; 3–Micrometric screw; 4–Threaded shaft sleeve;
5–Fixed sleeve; 6–Thimble; 7–Rachet; 8–Lock device.

Fig. 1-14 The configuration of the micrometer

2) The method of reading micrometer

The pitch of the right end of spindle is 0.5 mm. When the thimble rotates for a circle, the thimble moves by 0.5 mm. There are 50 grids on the circumference of the cone in front of the thimble, so when the thimble turns by one line, the spindle moves 0.01 mm.

The following steps show how to read the value of a micrometer:

(1) Read out the whole number of the size of the fixed sleeve behind the thimble.

(2) Count which line on the fixed is aligned with the reference line on the thimble.

(3) Add the two numbers together to get the final size, as shown in Fig. 1-15.

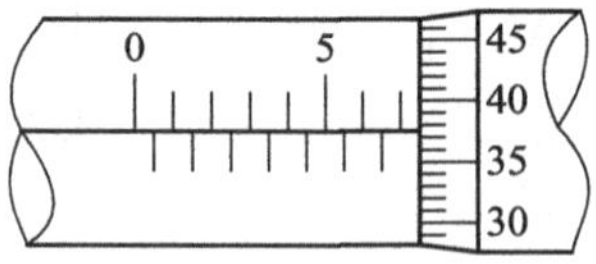

Fig. 1-15 The value reading of a micrometer

3) The use methods of the micrometer

Before measuring with a micrometer, clean the measuring surface of anvil and micrometric screw, and calibrate the zero position of micrometer, as shown in Fig. 1-16. The measurement can be operated with one hand or both hands, and the specific method is shown in Fig. 1-17. No matter which method is used, the rotation force should be appropriate. Generally, the thimble should be rotated first. When the measuring surface are about to contacts or have just touched the surface of the workpiece, then the ratchet should be rotated to control a certain measuring force, and finally read the value.

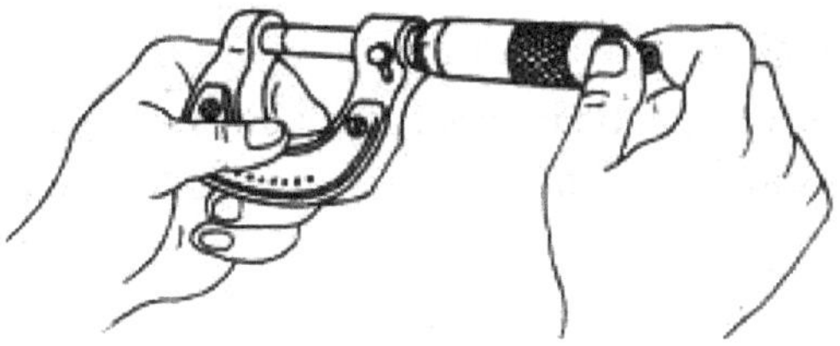

Fig. 1-16 Zero position inspection of micrometer

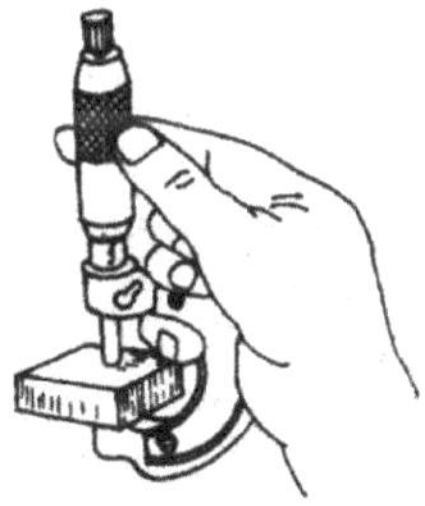

(a) Measuring with one hand

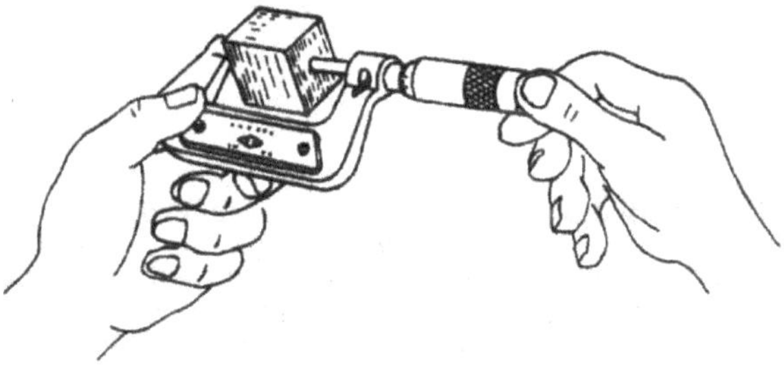

(b) Measuring with both hands

Fig. 1-17 The use methods of the micrometer

The following should be paid attention to when using the micrometer:

(1) The measuring surfaces of micrometer should be kept clean, and it should be calibrated before use.

(2) When measuring, first turn the thimle, and when the measuring surfaces get close to the workpiece, use ratchet instead until the ratchet squeaks.

(3) It is not allowed to measure the workblank with micrometer, let alone measure workpieces when they are still rotating.

(4) After the measurement, the micrometer should be kept clean, and a certain gap should be kept between the two measuring surfaces of the micrometer with the measuring range of 0–25 mm.

IV. Repair and maintenance of measuring tools

The measuring tools not only must be used correctly and reasonably, but also be properly repaired and maintained to avoid premature lose accuracy or damage to the measuring tool. Therefore, the following points should be done in use:

(1) Measuring tools(especially accuracy measuring tools)should be checked and maintained regularly. In case of any abnormal phenomenon, the user shall send it to the metering room for maintenance in time.

(2) The parts of the measuring tool should be complete, and the measurement should not be carried out in case of missing parts, so as not to affect the measurement accuracy.

(3) Before measuring, the working surfaces of the measuring tool and the measured surfaces of the workpiece should be cleaned to avoid the influence of dirt on the measurement accuracy or acceleration of the wear of the measuring tool.

(4) Do not stack measuring tools together with knives and other tool, so as to avoid scratches, damage, or deformation by squeezing.

(5) After measuring, clean all parts in time, apply oil and store it in a special measuring tool box to isolate the magnet and prevent deformation. It should be kept dry to avoid rust.

Task Four Cognition of dimensional tolerance

※ Knowledge objectives and skill objectives

(1) Understand the concept of dimensional tolerance.

(2) Understand the grade code of dimensional tolerance.

(3) Be able to read the dimensional tolerance in drawings.

In the process of machining mechanical parts, various errors will inevitably occur. The dimensional error directly determines the matching property of two matching parts. Only when the dimensional error of the parts is controlled within the allowable range, the interchangeability requirement can be met.

Dimensional tolerance is used to limit the manufacturing dimensional error of parts. It represents the size of tolerance zone and affects the fitting accuracy. Whether the machined parts meet the tolerance requirements, it needs to be judged by inspection. Inspection is not only used to judge whether the parts are qualified or not, but also to analyze the causes of unqualified parts, so as to timely adjust the production, supervise the process and avoid waste products. The following concepts need to be mastered during inspecting.

1. Size

Size is a value that represents linear length in a specific unit. The specific unit refers to the situation that in mechanical drawing, the sizes in the drawing are usually recorded in millimeters. When using millimeters as the unit, it is allowed to mark only the number and omit the unit when marking; when the drawing is not recorded in the millimeter, the number and unit must be marked completely.

2. Nominal size

Nominal size refers to the basic size of ideal shape determined by drawing specifications. The nominal size can be a whole number or a decimal value. The limit size can be calculated by nominal size and the upper and lower limit deviations.

3. Limit of size

The limit of size refers to the two limit values that allow for size variation. The limit of size is determined according to the design requirements, and its purpose is to limit the variation range of the actual size. Between the two sizes, the larger one is called the maximum limit of size "max", and the smaller one is called the minimum limit of size "min". The limit of size can be bigger than, equal to or less than the nominal size.

4. Dimensional tolerance

Dimensional tolerance refers to the difference between the high limit dimension and the low limit dimension, or the difference between the high limit deviation and the low limit deviation. It is the variation of the allowable dimension. Tolerance represents the variation range of a dimension, which is an absolute value and in front of it, no symbol should be added.

5. Tolerance zone of size

Tolerance zone of size refers to a zone defined by two straight lines representing high limit deviation and lower limit deviation or high limit dimension and lower limit dimension, or an area defined by one or several ideal geometric lines or surfaces, whose size is represented by linear tolerance value. It is determined by the size of the tolerance and its position relative to the zero line, as shown in Fig. 1-18.

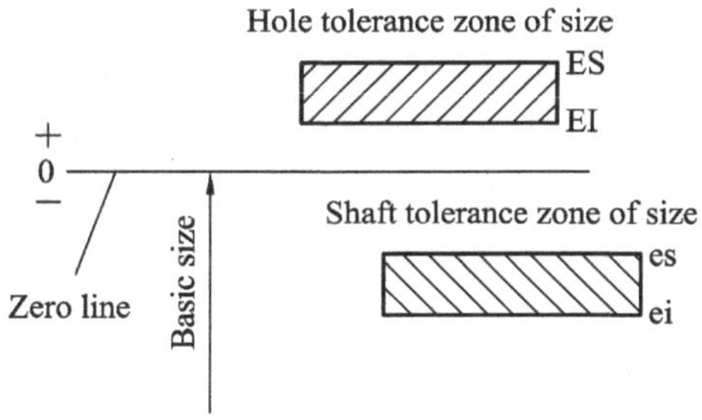

Fig. 1-18 Tolerance zone of size

(1) High limit deviation code. As specified, about the high limit deviation code, the hole is represented by uppercase letters "ES" and the shaft is represented by lowercase letters "es".

(2) Low limit deviation code. As specified, about the low limit deviation code, the opposite holes is represented by uppercase letters "EI" and the shaft is represented by lowercase letters "ei".

6. Standard tolerance and grade code

Standard tolerance grade code consists of symbol "IT" and number. When it forms a tolerance zone of size with the letter representing the basic deviation, the letter "IT" is omitted.

There are 20 standard tolerance grades of IT01, IT0, IT1–IT18 for the nominal size within 0–50 mm and 18 standard tolerance grades of IT1–IT18 for the nominal size within 500–3150 mm. The standard tolerance values of IT1–IT5 with nominal size greater than 500mm are in trial use. When the nominal size is less than or equal to 1 mm, there is no IT4–IT18. A set of tolerances of the same tolerance grade for nominal size is considered to have the same degree of accuracy. Different sizes and IT accuracy can be obtained by looking up the table.

Task Five Processing skills of simple parts

※ Knowledge objectives and skill objectives

(1) Learn the production process of parts.

(2) Learn the definition of the process flow and the process composition.

(3) Master the classification and selection principles of datum.

(4) Master the requirements, steps and methods of formulating process flow.

(5) Be able to formulate the process flow of simple parts.

I. Production process

In the manufacturing process, due to the different requirements and production conditions of parts, the manufacturing procedure is different. If the same parts are produced according to different processing plans, their production efficiency and economic benefits are also different. On the premise of ensuring the quality of parts, the process of drawing up reasonable and feasible process plan with good comprehensive technical and economic benefits is called the technological process design of parts.

From design drawings to products, a series of manufacturing processes are required. Generally speaking, the whole process in which raw materials or semi-finished products are transformed into products is called production process. The production process usually includes:

(1) Technical preparation process: including market investigation, prediction, new product identification, process design, standardization check, etc.

(2) Technological process: refers to the process of directly changing the size, shape, mutual position of the surface, surface roughness or performance of the raw material or semi-finished products to make them finished products. For example, liquid forming, plastic forming, welding, powder forming, cutting, heat treatment, surface treatment, assembly and so on, all belong to the process. When compiling the reasonable process into a technical document to guide production, the document is a process specification.

(3) Auxiliary production process: refers to the auxiliary production activities necessary to ensure the normal operation of the basic production process.

(4) Service process: refers to the organization, transportation, storage, supply, packaging and sales of raw materials.

The technological process mentioned above includes many procedures, and each procedure is composed of working position, working step, feed and installation.

Working procedure refers to the part of the technological process that is continuously completed for one or a group of workpieces on a machine tool or at the same working place. The basis of dividing working procedure is whether the working place changes and whether the work is continuous.

◇ Working position: the fixed part relative to the tool or equipment. Each processing position occupied by the workpiece is the working position.

◇ Working step: refers to the part of the working procedure that is completed when the machining surface of the workpiece, the cutting tool and the rotational speed and feed rate in the cutting parameters remain unchanged. When there are many processes for machining surface, it can be divided into several steps. Working step is the basic unit of working procedure.

◇ Feed: when the tool cuts off a layer of margin from the machined surface, it is called the feed. When the margin to be cut is too large to be cut at one time, it can be cut little by little, finished in several times.

◇ Installation: the part of the procedure completed after the workpiece is clamped once (positioning and clamping) is called installation.

In most cases, the workpiece is only installed once in a process, and sometimes it may be installed many times.

II. Datum and Positioning

1. Definition and classification of datum

Datum refers to the point, line and surface on which the position of some points, lines and surfaces on the part (or component) is determined. That is, datum is the point, line or surface of the part itself or related to the part. According to these points, lines or surfaces, the position of other points, lines or surfaces on the part is determined.

According to their different functions, datum can be divided into design datum and process datum. The process datum can be divided into positioning datum, measuring datum and installation datum, and the positioning datum can be subdivided into rough datum and fine datum.

(1) Positioning datum: refers to the datum used for positioning the workpiece in the machine tool or fixtures during processing.

(2) Measuring datum: refers to the datum used to measure the size and position of the machined surface during part inspection.

(3) Installation datum: refers to the datum used to determine the position of parts in components or products during installation.

2. Principles of datum selection

1) Selection of rough datum

In order to make all machined surfaces have enough machining allowance and ensure that each machined surface has a certain position accuracy to the unmachined surface, the selection of rough datum should follows the following principles:

(1) Select the unmachined surface on the workpiece as the rough datum.

(2) Select the important surface or the surface with the minimum machining allowance as the rough datum.

(3) Select a relatively smooth and large surface as the rough datum.

(4) Generally, the rough datum can only be used once, and try to avoid repeated use.

2) Selection of fine datum

The following principles should be abided by when selecting the fine datum:

(1) The principle of datum coincidence. That is, choose the design datum as the positioning datum as far as possible, so as to avoid the positioning error caused by the misalignment of the datum.

(2) The principle of the same datum. The same accuracy datum should be used for as many surface processing as possible.

(3) General requirements. When selecting a fine datum, accurate positioning, stability and reliability are required, and the fixtures can be simple and easy to operate.

3. Positioning of the workpiece

1) Six-point positioning principle

Any rigid body has six degrees of freedom in the space rectangular coordinate system. In order to get the accurate position of the workpiece in the fixture, the corresponding degree of freedom must be restricted.

2) Several situations in positioning

(1) Complete positioning: the positioning in which all six degrees of freedom of the workpiece are restricted.

(2) Incomplete positioning: the positioning in which, according to the processing requirements of the workpiece, it is not necessary to restrict the positioning of all degrees of freedom of the workpiece.

(3) Over positioning: the positioning in which, two or more positioning elements repeatedly restrict the positioning of the workpiece with the same degree of freedom.

(4) Under positioning: the positioning in which, according to the processing requirements of the workpiece, the degree of freedom that should be restricted is not completely restricted.

III. Drafting of process procedures

Process procedures is a technical document to guide production. The process procedures must be economical and reasonable.

1. Requirements and steps for formulating process procedures

The process procedures of parts are the processing methods and steps of parts, including the processing technology (heat treatment process), the determination of the machine tool, installation and clamping method, measuring method, machining allowance, cutting amount and the working hour quota used in each process. To fill in various contents on a certain form of card is the procedure of machining process, that is, the machining process card.

1) Requirements for formulating process procedures

Different parts have different structure, size, accuracy and surface roughness requirements, so the processing technology is also different. There may be several process plans for a part, and the most reasonable one should be selected. Reasonable processing technology must be able to ensure all the technical requirements of the parts. It is not easy to formulate a reasonable processing technology. It often requires repeated practice and modification for gradual improvement.

2) Steps of formulating process procedures

The steps of formulating process procedures are mainly as follows:

(1) make process analysis of parts.

(2) Select the workblank.

(3) Select the positioning datum.

(4) Formulate the process route.

(5) Select/design and manufacture machine tools.

(6) Select/design, manufacture cutting tools, fixtures, measuring tools and other auxiliary tools.

(7) Determine the machining allowance, process size and tolerance of the process.

(8) Determine the cutting amount of the process.

(9) Estimate the time quota.

(10) Fill in the process documents.

2. Main problems to be solved in the process procedures formulating

1) Selection of workblank allowance

The choice of workblank has a great influence on economic benefit. Because the arrangement of working process, the consumption of materials and the amount of working hours, depend on the selected workblank allowance to a certain extent.

2) Process analysis of parts

The process analysis of parts mainly includes:

(1) Check whether the drawings of the parts are complete and correct, and analyze whether the accuracy, surface integrity and technical requirements of the parts' main surface can be achieved under the existing production conditions.

(2) Check whether the selection of part materials is appropriate and whether it will make the process difficult and complicated.

(3) Review the structural processibility of parts and check whether the structure of parts can be processed economically and effectively.

3) Selection of positioning datum

(1) Generally, the center holes at both ends are selected as the fine datum for the stepped shaft used for transmission.

(2) Generally, the center hole is used as the fine datum for disc and sleeve parts.

(3) Generally, for bracket box parts, the main plane on the machine base is used as the fine datum to process the supporting holes of the bearings.

4) Formulation of process route

(1) Determine the processing plan.

(2) Divide the processing stages.

(3) Arrange the processing sequence.

(4) Centralize and decentralize the process procedures.

(5) Determine the machining allowance.

(6) Fill in the process documents.

After the process route is drawn up, the process number, process content, process diagram, machine tools to be used and other items and contents should be filled into the technical documents in the form of charts.

IV. The processing technology analysis of shaft parts

Shaft parts are rotating parts, which are often encountered in the process of machining. According to the different structural shapes of shaft parts, it can be divided into hollow shaft, stepped shaft, optical shaft and crankshaft.The processing technology of shaft parts is analyzed as follows.

1. Workblanks and materials of shaft parts

1) Workblanks of shaft parts

Large shafts or shafts with complex structure are made of castings, commonly round bars and forgings. According to the different production scales, two forging methods are commonly employed: free forging and die forging.

2) Materials of shaft parts

The common materials of shaft parts include 45# steel, bearing steel, low-carbon alloy steel, spring steel 65Mn, etc.

2. General processing requirements and methods of shaft parts

1) Key points of processing procedures for shaft parts

Whether the processing procedures are reasonable or not will directly affect the labor productivity and economic benefits. The following points must be paid attention to when formulating mechanical processing procedures.

(1) Process analysis of parts drawings: This part mainly studies products assembly drawings, and makes preparation for technical requirements.

(2) Selection of positioning datum: Try to select the design datum or assembly datum as positioning datum, and make the positioning datum coincide with the measurement datum.

(3) Selection of rough datum: When selecting firm and reliable surface as rough datum, the unmachined surface should be selected as much as possible. At the same time, the rough datum

can not be used repeatedly.

(4) Formulation of process route: The processing sequence must be correct during the processing. Therefore, advanced processing methods should be used as much as possible to formulate reasonable process route.

2) Overview of processing technology for shaft parts

The processing methods of shaft parts are mainly introduced as follows:

(1) Using the turning tool for turning long and thin shafts. Generally, the rake angle and the entering angle of the turning tool are large, and the accuracy turning tool often has a certain negative edge inclination angle to reduce the radial vibration and bending deformation, so that the chips flow to the surface to be processed.

(2) Adopting reverse feed. The reverse feed is adopted to make the direction of the feed force exerted by the tool on the workpiece toward the tailstock, which greatly reduces the bending deformation caused by the elongation of the workpiece.

(3) Using a follow rest. Using a follow rest can balance out the influence of radial cutting force, and keep the center of the follow rest consistent with the very center of the machine tool, so as to reduce the cutting vibration and workpiece deformation.

(4) Improving the clamping method of the workpiece. In high-speed, large-volume cutting, there exists the danger of getting away from the center, which could be avoided by using the method.

3) Several problems on the processing sequence of shaft parts

After dealing with datum positioning, the following processing sequence problems should be minded:

(1) The processing of secondary surfaces such as milling splines and keyways are generally conducted after finishing turning the outer circle.

(2) Deep hole processing should be arranged after quenching and tempering, which can effectively avoid the influence of heat treatment deformation on the shape of the hole.

(3) The CNC machining equipment provides the basis for the modernization of production. CNC turning processing not only improves the machining accuracy, but also ensures the high efficiency of production.

(4) For the outer surface, the outer circle with larger diameter should be processed before the outer circle with smaller diameter.

3. Processing technology analysis of shaft parts

1) Processing route of shaft parts

The main processing routes of shaft parts are as follows:

(1) Rough turning→semi-finishing turning→finishing turning.

(2) Rough turning→semi-finishing turning→rough grinding→fine grinding.

(3) Rough turning→semi-finishing turning→finishing turning→diamond turning.

(4) Rough turning→semi-finishing→rough grinding→fine grinding→finishing.

2) Typical processing route

The typical processing route is: workblank and its heat treatment→pre-processing→turning the outer circle→milling keyway (spline groove, groove)→heat treatment→grinding→final inspection.

3) Positioning datum of shaft parts processing

The positioning datum of shaft parts mainly include the following points:

(1) Positioning with the center hole of the workpiece. The center hole is not only the positioning datum during turning, but also conforms to the principle of datum uniformity.

(2) The outer circle and the center hole are used as the positioning datum. This method can withstand larger cutting torque and is the most commonly used for shaft parts.

(3) The cone plug with a central hole is used as the positioning datum. The center hole on the cone plug and cone sleeve mandrel is not only the positioning datum for its own manufacture, but also the datum for the outer circle finishing of the hollow shaft. In production, the cone plug should not be removed or replaced after installation until the machining is completed.

(4) The two outer circular surfaces are used as the positioning datum. This positioning datum can eliminate errors caused by the datum misalignment. When machining the inner hole of the hollow shaft, the two outer circular surfaces of the shaft can be used as the positioning datum.

Questions for review

1. What is the definition of the fitter and what does a fitter do?
2. What does the workshop 5S management refer to?
3. What are the common equipment and measuring tools for the fitter?
4. How is the three-view drawing of basic geometry read?
5. What is the purpose of steel heat treatment?
6. Try to analyze the marking principle and reading method of vernier caliper with a division value of 0.02 mm.
7. What are the classification and selection principles of datum?
8. Briefly describe the process of shaft parts.

Module Two Parts processing operation

Task One Marking

※ Knowledge objectives and skill objectives

(1) Understand the definition of marking and its role for the fitter.

(2) Master the selection of marking datum.

(3) Use the plane marking tools correctly.

(4) Master the general marking method, and be able to sample and punch holes on the line strip correctly.

I. Basics of Marking

(I) Definition of Marking

1. Definition of marking

Marking refers to use marking tools to draw the contour lines or points and lines of the parts to be processed on the workblank or workpiece, based on the dinerions of drawing or physical object as a reference. There are two kinds of marking: plane marking and three-dimensional marking. Plane marking is marking on a surface of the workpiece, which clearly shows the processing boundary. Only when several different surfaces of the workpiece are marked, the machining boundary can be reflected. This kind of marking is three-dimensional marking.

2. The function of marking

(1) Determine the machining position and machining allowance of each surface on the workpiece.

(2) Comprehensively check whether the shape and size of the workblank conform to the drawing and and whether it meets the processing requirements.

(3) When some defects appear on the workblank, the so-called “borrowing” method during marking can be used to remedy them in some way.

(4) Cut the material according to the marking line on the sheet, so the material can be arranged correctly and used reasonably.

(II) Paint for marking

1. Types of marking paint

In order to make the line clear, it is generally necessary to apply a thin and uniform coating on the marking part of the workpiece. Lime water is used on the rough surface of casting and forging workblanks (usually adding appropriate amount of oxhide gelatin to increase adhesion), and blue oil is used for processed surfaces.

2. Selection of marking paint

(1) Lime water: It is suitable for the marking of general castings and forgings. Lime water can be got by soaking the hydrated lime into water. Appropriate amount of oxhide gelatin can be added to increase the adhesion.

(2) Blue oil: It is made up of gentian violet with mass fraction of 2%–4%, shellac paint with mass fraction of 3%–5% and alcohol with mass fraction of 91%–95%. It is suitable for marking the machined surface.

(III) Marking Datum

1. The concept of datum

Marking datum refers to the basis used to determine the position of other points, lines and surfaces on parts . The correct selection of marking datum is the key to marking operation. Only with reasonable datum , Marking can be accurate, convenient and efficient.

2. Determination of the datum line of plane marking

1) Datum form of plane marking

(1) Take two mutually perpendicular planes (or straight lines) as the datum.

(2) Take two mutually perpendicular center lines as the datum.

(3) Take a plane and a center line as the datum.

2) Determination of datum line

The marking datum should be consistent with the design datum. When marking, one must start from the datum line, that is to say, determine the position of the datum line first, and then draw the position lines and shape lines of other surfaces. The drawing is the basis of marking. Before marking, the drawing must be carefully analyzed to ensure the marking process is correct. Only in this way, the processing limit of parts can be drawn according to the requirements of the drawing.

3. The selection principle of marking datum

(1) The marking datum should coincide with the design datum as far as possible.

(2) For the workpiece with symmetrical shapes, the symmetrical center line should be taken as the datum.

(3) For the workpiece with holes or bosses, the center line of the main hole or boss should be taken as the datum.

(4) When marking on the unprocessed workblank, the unprocessed surface should be taken as the datum.

(5) When marking on the machined workpiece, the machined surface should be taken as the datum.

II. Common tools and usage for marking

In marking, in order to ensure the accuracy of the size and achieve higher efficiency, worker must be familiar with all kinds of marking tools and correct usage.

(1) Steel ruler is a simple measuring tool. It is mainly used to measure the size, measure the workpiece, and can also be used as a guide tool for marking.

(2) The markinging plate is also called the marking platform, as shown in Fig. 2-1. It is made of cast iron. The surface of marking plate is processed by finish planing or scraping. The marking plate is generally placed on a wooden stand, and is in a horizontal state.

Since the surface of the marking plate is the basic plane for marking, its flatness directly affects the marking, so the working plane must be kept in a horizontal position during installation. During use, the surface of the marking plate should be kept clean to prevent iron filings, lime sand, etc. from scratching the surface when the marking tool or workpiece moves. Workpieces and tools should be handled with care, so as not to damage the working surface. It is not allowed to strike on the marking plate. The marking plate should be used evenly to avoid local bulges so as to affect the flatness of the marking plate. After use, it should be wiped clean and coated with oil to prevent rust.

Fig. 2-1 Marking plate

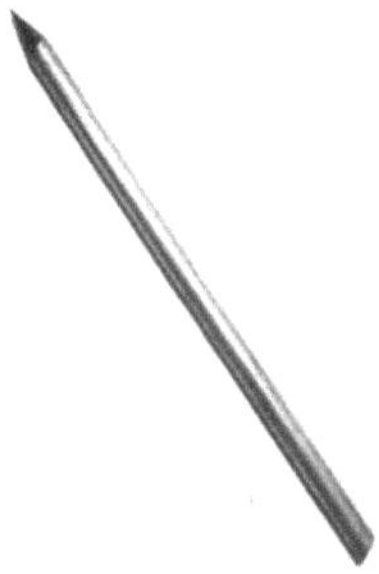

Fig. 2-2 Marking pin

Fig. 2-3 Marking block

(3) The marking pin (see Fig. 2-2) is used to draw lines on the workpiece. When marking, it is generally used together with guiding tools such as steel rulers squares or templates. The striking needle is made of spring steel wire or high-speed steel, with a diameter of 3–5 mm in general, and the tip is ground into a sharp angle of 15°–20°. Some have cemented carbide welded on the tip for better wear resistance.

(4) The marking block (see Fig. 2-3) is used to mark the workpiece on the marking platform or to align the position of the workpiece on the platform. The straight end of the marking block is

used for marking, and the elbow end is used for aligning the position of the workpiece.

When marking, the marking pin should be placed in a horizontal position as much as possible, not too inclined. The extended part of the marking pin should be as short as possible and clamped firmly to avoid vibration and change. When drawing a long straight line, the segmented connection method should be adopted to facilitate the end-to-end check. After the use of the marking block, the marking pin should be kept in a horizontal state to ensure safety and reduce the space occupied.

(5) The height vernier caliper (see Fig. 2-4) is equipped with a marking pin, which can directly indicate the height dimension. Generally speaking, It's reading accuracy is 0.02 mm, which can be used as a precision marking tool.

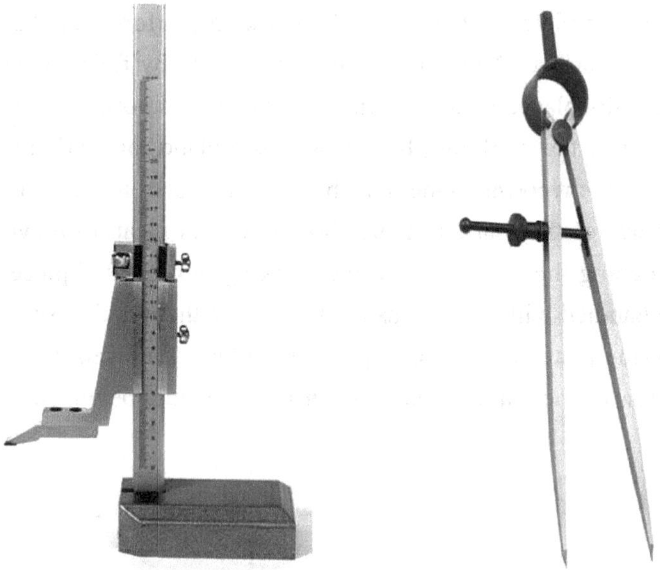

Fig. 2-4 Height vernier caliper and scribing compass

(6) The scribing compass (see Fig. 2-4) is used to draw circles, arcs, bisection lines, bisection angles and measure sizes. There are common scribing compass, spring scribing compass and long scribing compass for the fitter. Generally speaking, the scribing compass is made of tool steel, and the pin is quenched. Some kinds of scribing compass are welded with cemented carbide on the pin to make it sharper and more wear-resistant.

(7) The sample punch is used to punch holes (punch points) on the processing lines of the workpiece, which is used to strengthen the boundary mark and to locate the center when drawing an arc or drilling. It is generally made of tool steel and hardened at the tip. The top angle is about 40° when used to strengthen the boundary mark and about 60° when used to drill and locate the center.

(8) The 90°square is often used as a guide tool for drawing parallel or vertical lines when marking, and can also be used to align the vertical position of the workpiece plane on the platform.

III. Plane marking practice

The drawing of the plane marking is shown in Fig. 2-5.

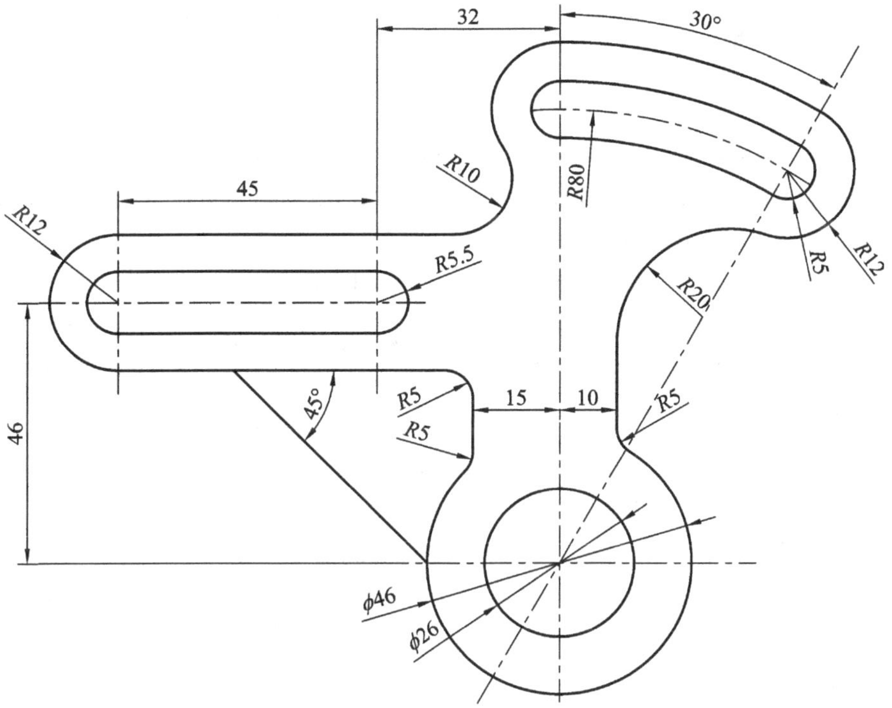

Fig. 2-5 Plane marking

(I) Preparation before operation

(1) Preparation of machining tools and measuring tools. Prepare the machining tools and measuring tools for the parts to be processed according to the drawings (see Table 2-1).

(2) Workpiece cleaning. Remove pouring risers, flashes, burrs, oxide scales, etc. on the castings and forgings.

(3) Coloring of the workpiece. In order to make the line clearer, before marking, apply a layer of uniform coating on the marking part of the surface of the unprocessed workpiece. Commonly used paints are chalk, lime water, blue oil or copper sulfate solution, etc. Chalk is used for small quantity and small workpieces; lime water is used for casting and forging workpieces; blue oil or copper sulfate solution is used for processed workpieces. Some workpieces are not allowed to have scratches on the surface, and can not be painted. Their positioning and marking are often done on the machining center.

(4) Selection of machining tools and measuring tools. select machining tools and measuring tools properly according to the requirements of the drawing.

Table 2-1 Preparation list of machining tools and measuring tools

No.	Name	Specification	Quantity	Remarks
1	Steel ruler		1	
2	Scribing compass		1	
3	Marking pin		1	
4	Sample punch		1	
5	Hammer		1	

(5) Understand the scoring standard, analyze the sizes and other technical requirements. The scoring standard is shown in Table 2-2.

Table 2-2 Scoring standard

No.	Quality inspection content	Score allocation	Scoring standard	Actual measurement	Score
1	The color is thin and uniform	4 points	Overall evaluation		
2	The figure and its arrangement position are correct	12 points	3 points will be deducted for each mistake		
3	Clear lines and without overlapping lines	10 points	1 point will be deducted for each unclear line or overlapping line		
4	The tolerance of dimension and line position is ± 0.3 mm	26 points	2 points will be deducted for each place beyond tolerance		
5	Smooth connection of each arc	12 points	2 points will be deducted for each poor connection		
6	The tolerance of punching position 0.3 mm	16 points	2 points will be deducted for each offset punching		
7	Reasonable distribution of inspection sample punching	10 points	2 points will be deducted for each unreasonable distribution		
8	Correct use of tools and correct operation posture	10 points	2 points will be deducted for one unreasonable item		
Civilized and safe production		Score deduction	2 points will be deducted for each violation		
Site records:					

(II) Operation steps

(1) Prepare the marking tools to be used and clean the part for practice and color the marking surface.

(2) Check whether the workpiece to be marked has enough machining allowance.

(3) Familiar with the marking method of each drawing, and arrange the reasonable position of the drawing datum line on the part for practice according to the drawing datum and the maximum outline size.

(4) Complete the marking in sequence according to the drawing (no size is noted in the drawing, and the drawing line can be reserved).

(5) After rechecking and proofreading the figure and size, punch the inspection sample on the lines of the ϕ26 mm hole, the 45 mm long waist hole and the 30° arc waist hole in the drawing.

(III) Relevant precautions

(1) In order to be familiar with the drawing methods of various figures, one can practice on paper before marking.

(2) Master the way to use marking tools and how to mark.

(3) Make sure the the marking size is accurate, the line is clear, and the punching is correct.

(4) Place tools correctly.

(5) Any workpiece must be rechecked carefully after marking to avoid mistakes.

IV. Three-dimensional marking practice

The three-dimensional marking drawing is shown in Fig. 2-6.

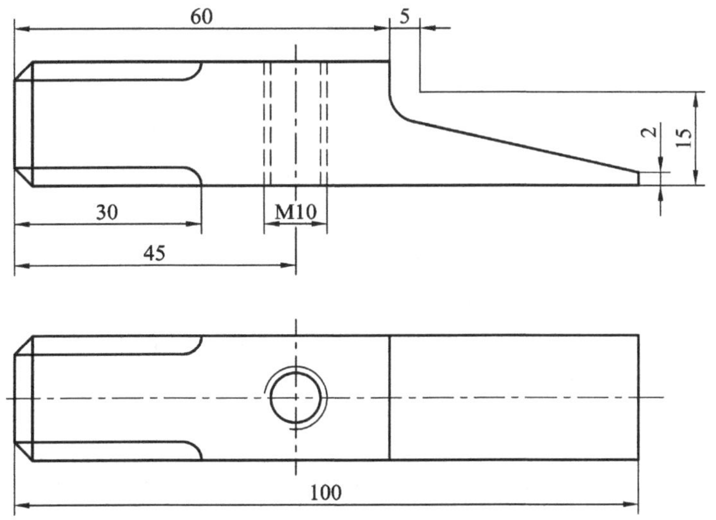

Fig. 2-6 The three-dimensional marking

(I) Preparation before operation

(1) Preparation of machining tools and measuring tools: prepare the machining tools and measuring tools for the parts to be processed according to the drawings (see Table 2-3).

(2) Workpiece cleaning: like plane marking practice.

(3) Coloring of the workpiece: like plane marking practice.

(4) Selection of machining tools and measuring tools: select machining tools and measuring tools properly according to the requirements of the drawings.

Table 2-3 Preparation list of machining tools and measuring tools

No.	Name	Specification	Quantity	Remarks
1	Height vernier caliper	0–300 mm	1	
2	Block		1	
3	Square box		1	
4	Steel ruler		1	
5	Scribing compass		1	
6	Marking pin		1	
7	Sample punch		1	
8	Hammer		1	

(5) Understand the scoring standard, analyze the sizes and other technical requirements. The scoring standard is shown in Table 2-2.

(II) Operation steps

(1) Prepare the marking tools to be used and clean the parts for practice and color the marking surface.

(2) Familiar with the drawing method of each drawing, and arrange the reasonable position of the drawing datum line on the practice piece according to the drawing datum and the maximum outline size.

(3) Complete the marking in sequence according to the drawing (no size is noted in the drawing, and the drawing line can be reserved).

(4) After rechecking and proofreading the figure and size, punch the inspection sample at the hole M10 and R5 in the drawing.

(III) Relevant precautions

The relevant precautions of the three-dimensional marking like plane marking practice.

Task Two Sawing

Sawing is a method of cutting or grooving materials or workpieces with a saw. Large-scale raw materials and workpieces are usually divided with a mechanical saw, which is not in the scope of a fitter's work. The sawing of a fitter refers to using a handsaw to divide or cut grooves on smaller materials or workpieces.

※ Knowledge objectives and skill objectives

(1) Understand the structure of handsaw.

(2) Know different kinds of saw blades and be able to clamp them correctly.

(3) Master the basic knowledge and process knowledge of sawing.

(4) Be able to saw materials of various shapes correctly, with the correct posture, and achieve certain sawing accuracy.

(5) Correctly select the saw blade according to different materials and know how to clamp.

I. Basic knowledge of sawing

(I) Handsaw

Handsaw is a hand tool used by the fitter for sawing operation, which is mainly composed of saw bow and saw blade. The saw bow is used to clamp and tension the saw blade. It has two types: the integral saw bow (as shown in Fig. 2-7(a)) and the split saw bow(as shown in Fig. 2-7(b)). The integral saw bow can only be installed with the saw blade of a certain length, while the split saw bow can be installed with saw blade of several lengths by adjustment. For the latter, the split saw bow is easy to work with force due to the shape of the saw handle, so it is widely used at present.

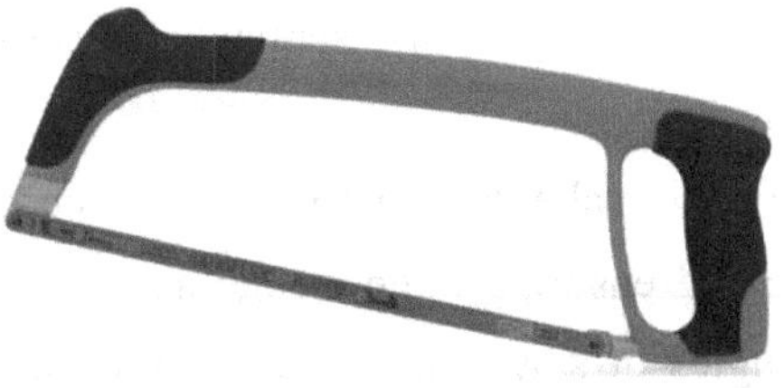

(a) Integral saw bow

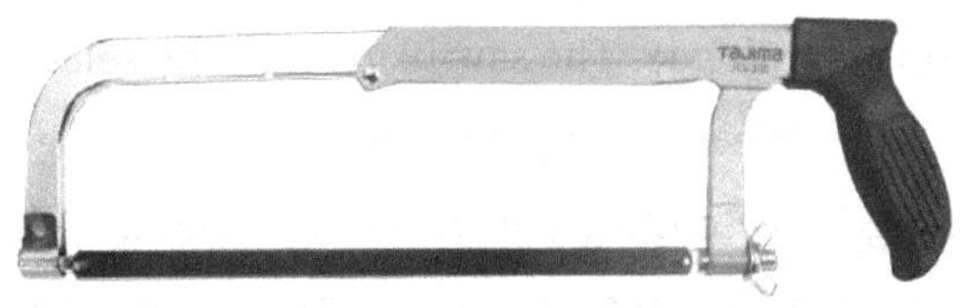

(b) Split saw bow

Fig. 2-7 Handsaw

(Ⅱ) Saw blade

1. The material of saw blade

The saw blade is usually made of high-quality carbon tool steel T10A or T12A. After heat treatment, the hardness can reach HRC60-64, which is the same as the material used for making files. Therefore, do not mix them together in normal operation, let alone stacking, so as to avoid relative friction and mutual damage. In addition, high-speed steel can also be used to make saw blades, which have higher hardness, toughness and heat resistance, but the cost is much higher than ordinary saw blades.

2. Saw blade specifications

The specifications of the saw blade mainly include two parameters: length and saw tooth.

(1) The length refers to the center distance of the installation holes at both ends of the saw blade, generally 100, 200, 300 mm and so on. The saw blades of 300 mm length, 12 mm width and 0.8 mm thickness are commonly used in the fitter practice.

(2) In order to adapt to the material properties and the width of sawing surface, the saw tooth is divided into three types: coarse, medium and fine. The coarse-tooth saw blade has a large tooth pitch and large chip space, which is suitable for sawing soft materials or workpieces with large saw sections. When sawing hard material, fine-tooth saw blade is selected. The thickness of the saw tooth is usually expressed by the number of teeth per 25 mm. The selection of the saw blade must be based on the comprehensive consideration of the thickness and hardness of the material at the sawing part (see Table 2-4).

Table 2-4 Classification and selection of saw blades

Categories	Tooth pitch /mm	Tooth number/25mm	Situations
Coarse tooth	>1.8	<14	The sawing part is thick and the material is soft
Medium tooth	1.1–1.8	14–22	—
Fine tooth	<1.1	>22	The sawing part is thin and the material is hard

3. Selection of the saw blade

The saw blade is the cutting part of a handsaw. Before sawing, it is essential to select the proper saw blade. The saw blade must be selected properly.

(1) When sawing, in order to achieve high efficiency and make the saw tooth have a certain strength, the cutting part must have enough chip groove and ensure a large wedge angle of the saw tooth. At present, the commonly used geometric angles of saw blades are as follows: the clearance angle α ranges between 40° and 50°, the wedge angle β ranges from 45° to 50° and the rake angle γ is 0°, as shown in Fig. 2-8.

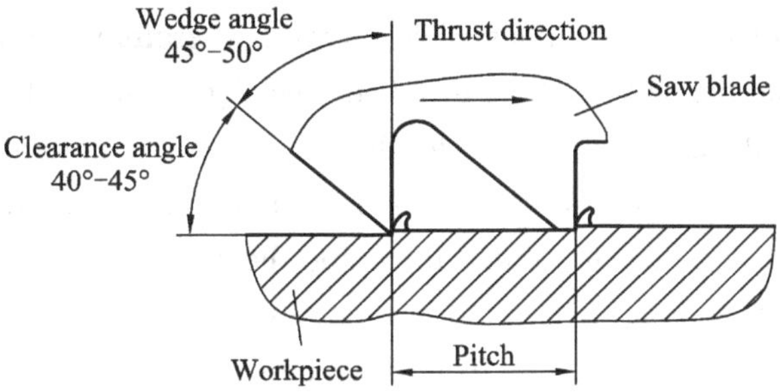

Fig. 2-8 Saw teeth angle

(2) When sawing, the deeper the workpiece is cut into, the greater the friction resistance of both sides of the saw slot to the saw blade; even the saw blade could be stuck. In order to avoid this situation, it is necessary to stagger the saw tooth on the saw blade according to a certain rule when manufacturing the saw blade. The saw tooth are arranged into a certain shape, which is called the kerf. The common saw kerf can be wavy or crossed, as shown in Fig. 2-9. The kerf is very important in the sawing process. Its existence prevents the two sides of the saw blade from directly contacting the workpiece, which will reduce the friction between the saw blade and the workpiece to reduce the heat generation and help to empty the iron filings. With the wear of both sides of the saw tooth, the saw path will become narrower and narrower, the resistance will become greater and greater, and the saw tooth will gradually lose their sharpness. When it reaches a certain level, the saw blade will lose its function of cutting.

(a) Wave-shaped (b) Cross-shaped

Fig. 2-9 Diagram of kerf

When sawing, the thickness of saw tooth should be selected according to the hardness of sawing material and the thickness of sawing surface. The coarse tooth saw blade is suitable for sawing soft materials and workpieces with large sawing surface. Because the soft material and the workpiece with larger cutting surface have more chips each time, and the coarse-tooth saw blade has a large chip groove, which will not cause blockage and affect cutting efficiency.

The fine tooth saw blade is suitable for sawing hard materials, which are not easy to be sawed in. It produces less chips so that the chip groove can not be blocked. Selecting the fine tooth saw blade can increase the number of teeth to be cut at the same time, so that the cutting amount of each tooth is reduced, the material is easy to be cut, the sawing is labor-saving, and the saw tooth is not easy to wear.

For workpieces with small sawing surfaces, such as pipes and thin plates, the fine tooth saw blade must be selected, otherwise the saw tooth are easy to be hooked and lead to tooth chipping.

4. Installation of the saw blade

The handsaw only works when it is pushed forward. Therefore, the direction of the tooth tip should be forward when the saw blade is installed (see Fig. 2-10(a)). If it is installed reversely (see Fig. 2-10(b)), the rake angle of the saw tooth is negative, and normal sawing is not possible. When adjusting the tension of the saw blade, the nut should not be too tight or too loose. If it is too tight, the saw blade will bear too much force and the saw blade will be broken easily when there is an improper use of force. If it is too loose, the saw blade is easy to twist and break, and the saw slot tends to skew. When judging its tension, one can pull the saw blade by hand to feel if it is hard and solid. After the saw blade is installed, the plane of the saw blade must be parallel to the central plane of the saw bow. The saw blade should not be tilted or twisted, otherwise, the saw slot will be easily skewed during sawing.

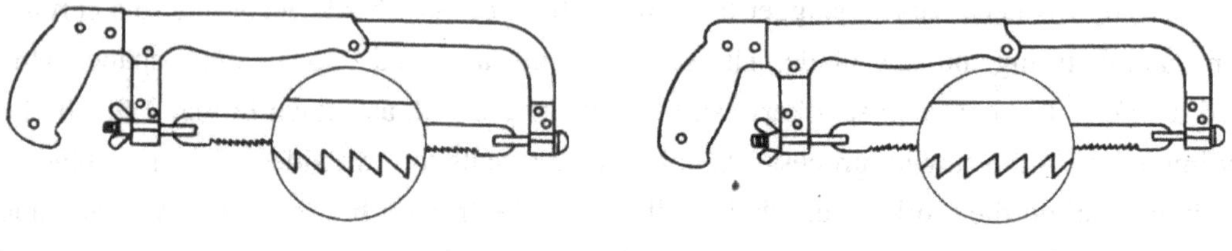

(a) Correct (b) Incorrect

Fig. 2-10 Saw blade installation

II. Key points of sawing

(I) Basic methods of sawing

(1) Grip method. Hold the handle with the right hand and hold the bow with the left hand, as shown in Fig. 2-11.

(2) Posture. The standing position and the body swinging posture when sawing are basically similar to filing, and swing naturally.

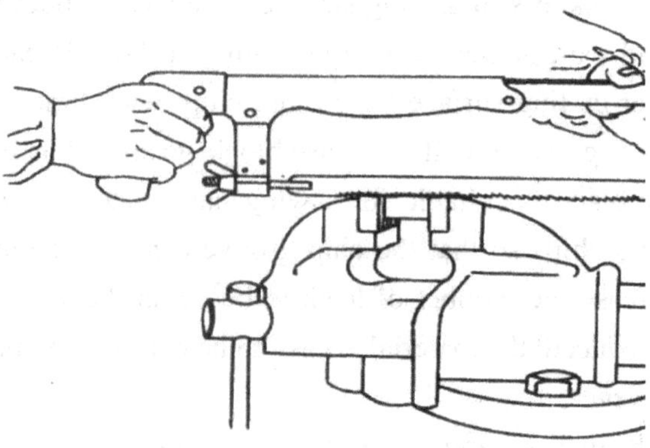

Fig. 2-11 The grip method of handsaw

(3) Pressure. When sawing, the thrust and pressure are controlled by the right hand, and the left hand mainly cooperates with the right hand to straighten the saw bow, and the pressure should not be too high. When the handsaw is pushed out, it is the cutting stroke, so the pressure should be applied. It does not cut during the return stroke, and it should be naturally pulled back without pressure. The pressure should be low when the workpiece is about to break.

(4) Movement and speed. Sawing movement generally adopts a small range of up and down swing movement. When the handsaw is pushed forward, the body slightly leans forward, with both hands pressed against the handsaw, the left hand tilted up, the right hand pressed down. During the return stroke/journey, the right hand is lifted up, and the left hand naturally follows back. It is required to cut the bottom surface of the saw seam straightly, and the straight-line movement must be adopted. The speed of sawing should be 20–40 times per minute. It should be slower to cut hard materials and faster to cut soft materials. At the same time, the speed of sawing stroke should be kept uniform, and the speed of return stroke should be relatively faster.

(II) Sawing operation methods

1. Workpiece clamping

Generally, the workpiece should be clamped on the left side of the bench vice for easy operation, and should not stretch out of the jaw too long (the saw slot should be about 20 mm away from the side of the jaw), so as to prevent the workpiece from vibration during sawing; the saw slot should be parallel to the side of the jaw (the saw kerf line should be aligned with the plumb line), so as to control the saw slot from deviating from the marking line; the clamping should be firm, and at the same time, avoid deformation of the workpiece clamping and damage to the machined surface.

2. Method of saw entry

Saw entry is the beginning of sawing. The quality of saw entry directly affects the overall quality of sawing. If saw enters improperly, the saw blade could jump out of the saw slot to roughen the workpiece or cause saw tooth chipping, and the saw slot after sawing is inconsistent with the marking position, which will lead to large deviation of sawing size.

There are two methods of saw entry: far saw entry [see Fig. 2-12(a)] and near saw entry [see Fig. 2-12(b)]. When starting sawing, put the left thumb against the saw back, so that the saw blade can be sawed in the required position correctly, with a short stroke, low pressure and slow speed. The sawing angle θ should be about 15°. If the entry angle is too large, it is not easy for the saw to start smoothly, especially when near saw entry is adopted, the saw tooth could be stuck by the edge of the workpiece and cause chipping [see Fig. 2-12(c)]. However, the entry angle should not be too small, otherwise, the large number of teeth in contact with the workpiece at the same time, it is not easy to cut into the material. It is easy to deviate after repeated sawing, which will cause many saw marks on the surface of the workpiece and affect the surface quality. In general, it is better to use far entry, during which the saw tooth cut into the material gradually, and accordingly they are not easy to get stuck, and it is more convenient to start. If the saw tooth get stuck by the edge of the workpiece due to poor technical skills, the handsaw can be pulled backward to start sawing reversely, so that the number of teeth contacted during entry increases, and then the saw tooth will not be stuck by the edge when thrusting to start sawing. When the saw blade reaches the depth of 2−3 mm, it will not slip out of the groove. Then the left thumb can leave the saw blade, and the saw bow can be straightened to make the saw mark level backward (forward) gradually, and then the saw blade will cut normally. During normal sawing, all the effective teeth of the saw blade should participate in cutting in each stroke.

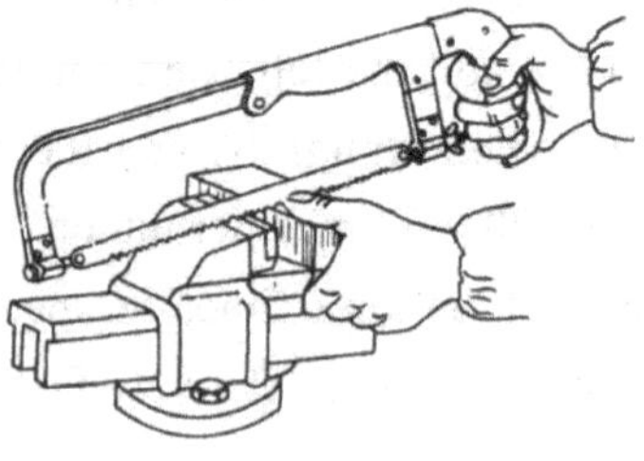

(a) Far saw entry

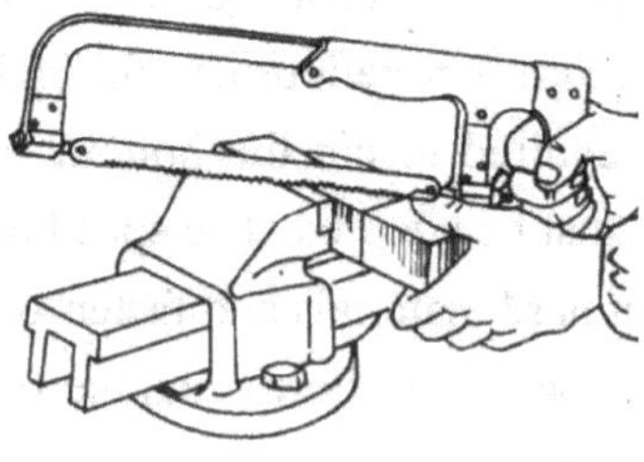

(b) Near saw entry

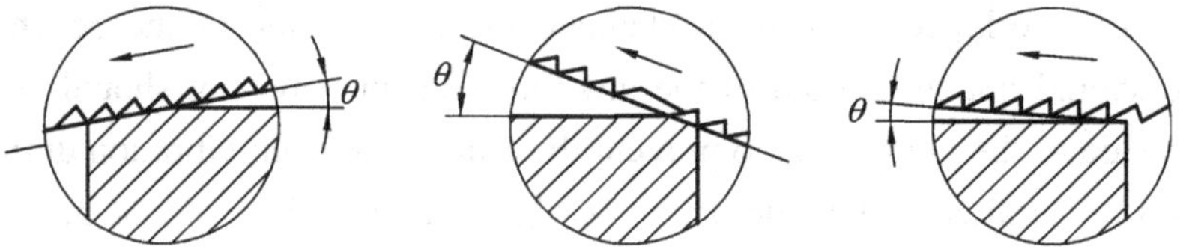

(c) The angle of saw entry

Fig. 2-12 Methods of saw entry

(III) Sawing methods of several materials

1. Sawing of bar

When sawing the bar, if the sawing section is required to be smooth, it should be sawed continuously from the beginning to the end. If the sawing section doe not have special requirements, it can be sawed in several directions. In this way, the sawing surface becomes smaller and it becomes easier to cut in, which can improve work efficiency.

2. Sawing of pipe

Before sawing a pipe, a sawing line perpendicular to the axis should be drawn. Since sawing does not require high accuracy for the marking, a rectangular paper strip (the marking side must be straight) can be used to surround the outer circle of the workpiece according to the sawing size, and then mark out with talc. The pipe must be clamped upright when sawing. For thin-walled pipes or finished pipes, they should be clamped between two wooden gaskets with V-shaped grooves [see Fig. 2-13(a)] to prevent flattening the pipe or damaging the surface.

When sawing a thin-walled pipe, it is not allowed to cut continuously from the beginning to the end in one direction, otherwise the saw tooth are easily hooked by the pipe wall and cracked. It is right to saw to the inner wall of the pipe in one direction, and then turn the pipe to a certain angle in the direction of the saw thrusting, connect the original saw slot and then saw to the inner wall of the pipe. In this way, gradually change the direction and continuously turn the saw until it is cut through [see Fig. 2-13(b)]

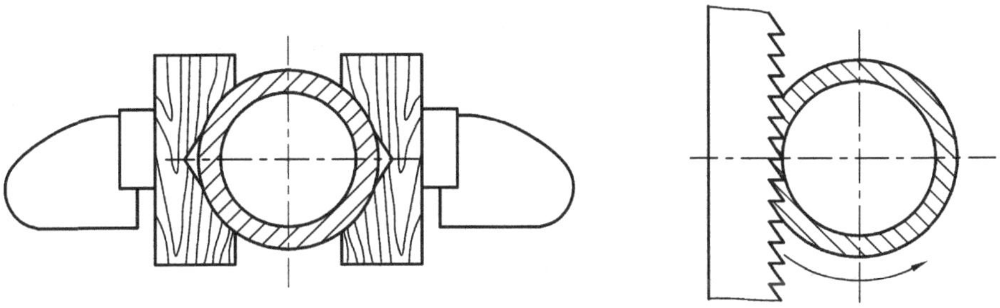

(a) Pipe clamping (b) Sawing while changing the direction

Fig. 2-13 Pipe clamping and sawing

3. Sawing of thin sheet

When sawing, try to cut from the wide surface. When it can only be sawed from the narrow surface of the sheet, it can be clamped by two pieces of wood, and the wood can be sawed off with the wood block, so as to avoid the hooking of the saw tooth. At the same time , the stiffness of the sheet is increased, so that no vibration occurs during sawing [see Fig. 2-14(a)]. The thin sheet can also be directly clamped on a bench vice, and a handsaw can be used to make a lateral oblique thrust, so as to increase the number of teeth contacting with the sheet and avoid the saw tooth chipping [see Fig. 2-14(b)].

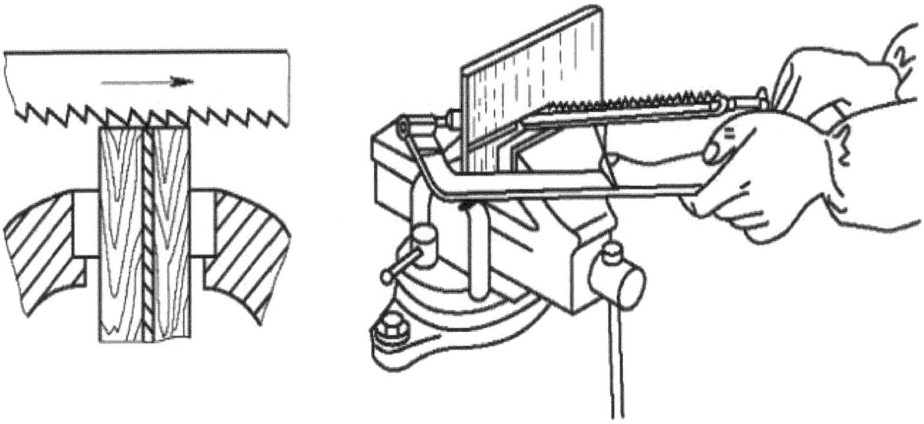

(a) (b)

Fig. 2-14 Sawing method of thin sheet

4. Deep slot sawing

When the depth of the saw slot exceeds the height of the saw bow [see Fig. 2-15(a)], the saw blade should be rotated by 90° to make the saw bow turn to the side of the workpiece [see Fig. 2-15(b)]. When the height of the saw bow is still not enough, the saw blade can also be clamped to make the saw tooth face inside the saw for sawing [see Fig. 2-15(c)].

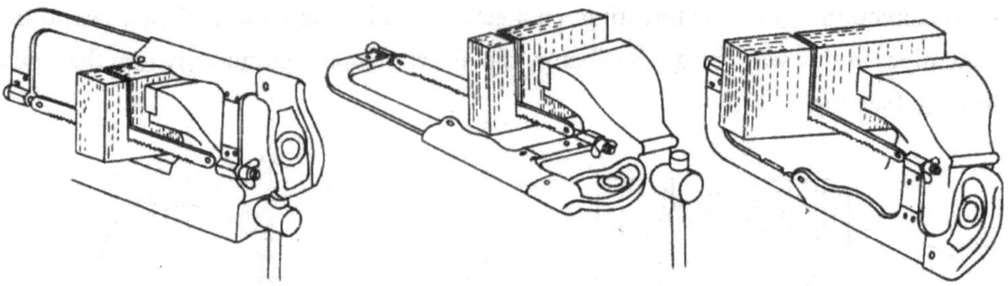

Fig. 2-15 Sawing method of deep slot

(IV) Safety knowledge of sawing operation

The safety knowledge of sowing operation are as follows:

(1) The saw blade should be installed with proper tension. When sawing, do not use too much force suddenly to prevent the saw blade from breaking off from the saw bow and hurting people.

(2) When the workpiece is about to be sawed off, the pressure should be small to avoid the accident caused by the sudden disconnection of the workpiece due to excessive pressure and the forward rush of the hand. When the workpiece is about to be sawed off, hold the broken part of the workpiece with the left hand to avoid foot injury caused by the workpiece falling.

(3) Handleless handsaw or straight handle handsaw without hoops should not be used to prevent the tail tip from stabbing the palm.

III. Sawing operation practice

The $\phi34\times120$ 45# steel as shown in Fig. 2-16, please practice sawing squares (as shown in Fig. 2-17) in its length direction.

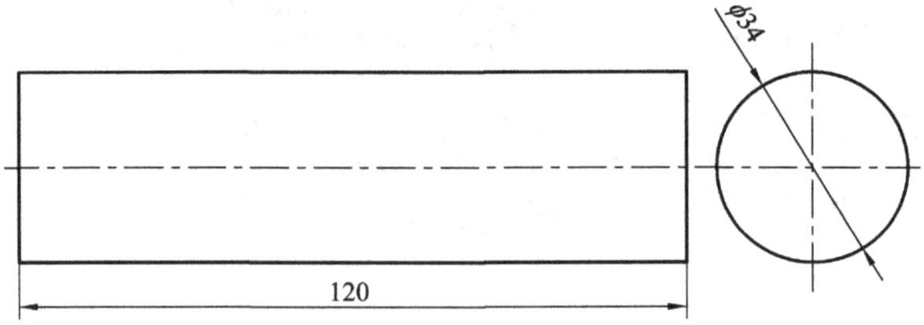

Fig. 2-16 Workblank drawing

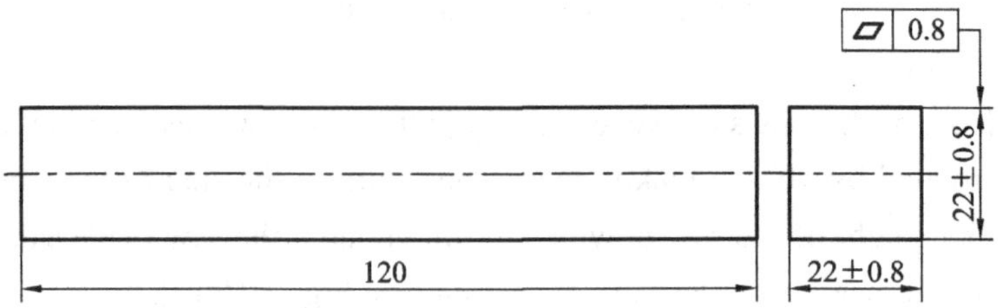

Fig. 2-17 Parts drawing

(Ⅰ) Preparation before operation

(1) Analyze drawings and clarify requirements: analyze and understand the parts drawing. This drawing requires that (22±0.8)mm (2 places) and flatness (4 places) should be ensured, and the two planes should be parallel to each other.

(2) Select machining tools and measuring tools: according to the drawing requirements, choose the machining tools and measuring tools properly. See Table 2-5 for details.

Table 2-5 List of machining tools and measuring tools

No.	Name	Specification	Quantity	Remarks
1	Vernier caliper	0–150 mm	1	Precision of 0.02 mm
2	90° square		1	
3	Steel ruler		1	
4	Bench vice		1	
5	Saw bow		1	
6	Saw blade		A lot	
7	Marking pin		1	
8	Brush		As required	

(3) Understand the scoring standards: analyze the sizes and other technical requirements, understand the key points of scoring and the important processing points. The scoring standard is shown in Table 2-6.

Table 2-6 Scoring standard

No.	Quality test content	Score distribution	Evaluation standard	Actual measurement	Score
1	(22±0.8) mm (2 places)	2 × 20	No score if beyond the tolerance		
2	flatness 0.8 mm (4 places)	4 × 10	No score if beyond the tolerance		
3	Safe and civilized production	20	No score for violations		

(II) Operation steps

The machining process should be developed and the operation should be carried out based on the process. The specific operation steps are as follows:

(1) Analyze the drawings, and check the workblank size.

(2) Mark the sawing line on the workblank according to the size of the drawings.

(3) Saw four planes separately. The requirements of (22±0.8) mm in size and 0.8 mm in

flatness of sawing section should be met and ensure that the saw marks are neat.

(4) All inspections should be carried out according to the drawings, and the burrs should be properly trimmed before the workpieces are handed over.

(Ⅲ) Relevant precautions

(1) During the sawing practice, one must check whether the installation of the workpiece and the saw blade are correct, and whether the saw entry method and the entry angle are correct, so as not to cause waste products and saw blade damage at the beginning of sawing.

(2) For beginners in sawing, it is not easy to master the sawing speed, and the thrusting speed is often too fast, which will blunt the saw blade quickly. At the same time, there are wrong postures such as unnatural swing and excessive swing, which should be corrected in time.

(3) It is necessary to pay attention to the straightness of the saw slot and correct it in time (The quality of sawing can not be guaranteed, if the correction is done when there is too much deviation).

(4) When sawing steel parts, some engine oil can be added to reduce the friction between the saw blade and the sawing section and cool the saw blade, which can improve the service life of the saw blade.

(5) After sawing, the butterfly nut on the saw bow should be properly loosened, but the saw blade should not be removed to prevent the parts on the saw bow from being scattered. They should be placed properly.

Task Three Filing

Filing is a method of cutting a workpiece with a file. The dimension accuracy of the file can reach about 0.01 mm, and the minimum surface roughness value can reach about $Ra0.8$ μm. Filing has a wide range of work. It can file the internal and external surfaces and various grooves of the workpiece. The fitter often uses all kinds of files to trim parts during the assembly process.

※ Knowledge objectives and skill objectives

(1) Understand the material, structure, type and selection principle of the file.

(2) Understand the maintenance of file and the safety knowledge when filing.

(3) Master the filing posture and the method of exerting force by both hands.

(4) Master the method of plane filing, and correctly grasp the filing speed.

(5) Master the measuring tools and methods of flatness, perpendicularity and parallelism.

(6) Master the knowledge of safe and civilized production of filing.

I. The Cognition of the File

(I) Structure of the file

File is the main tool for filing. It is commonly made of carbon tool steel T12 and T13 and hardened to HRC62-67 through heat treatment. The file is brittle and easy to break, so it should be carefully protected during use.

The file is composed of file body (including the file face and the file edge) and the file handle, whose structure is shown in Fig. 2-18. The file face is the main working face of file. Generally, the front end of the file face is made into a convex arc shape, which is convenient for filing part of the workpiece plane. The file edge refers to the two sides of the file, some of which have teeth on one side and no teeth on the other side (called the smooth edge), so that when filing the workpiece with inner right angle, the other adjacent surface can be protected. The file tang is used to hold the file handle.

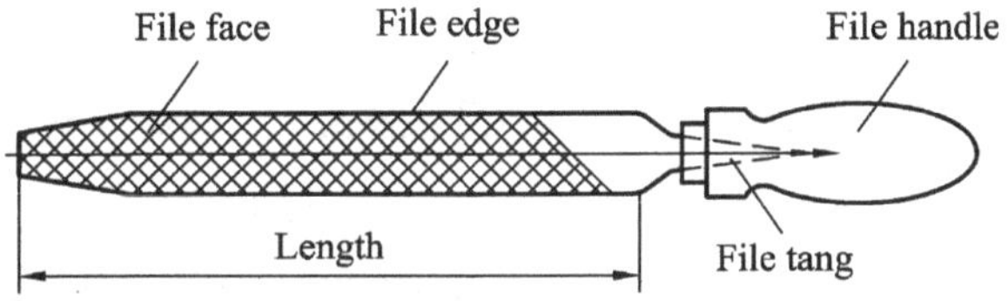

Fig. 2-18 The structure of the file

The tooth pattern of a file can be divided into two kinds: single tooth pattern and double tooth pattern. Generally, the edge of the file is made into single tooth pattern, and the face of the file is made into double tooth pattern. The bottom tooth angle is 45° and the face tooth angle is 65°, as shown in Fig. 2-19.

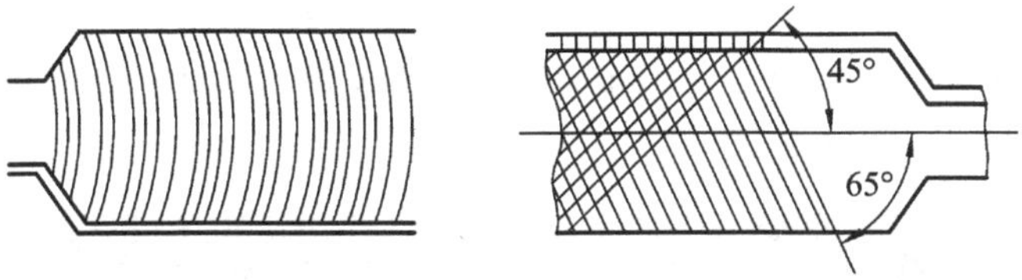

(a) Single tooth pattern (b) Double tooth pattern

Fig. 2-19 The tooth pattern of file

(II) Types and application of the file

According to different purposes, files can be divided into several types: fitter files, shaping files and irregular files, etc. The file commonly used by the fitter is the fitter file.

The fitter file can be classified as flat file (plate file), half round file, square file, triangular file and oval file according to different cross-section shapes, as shown in Fig. 2-20. The fitter file and their suitable processing surfaces are shown in Fig. 2-21.

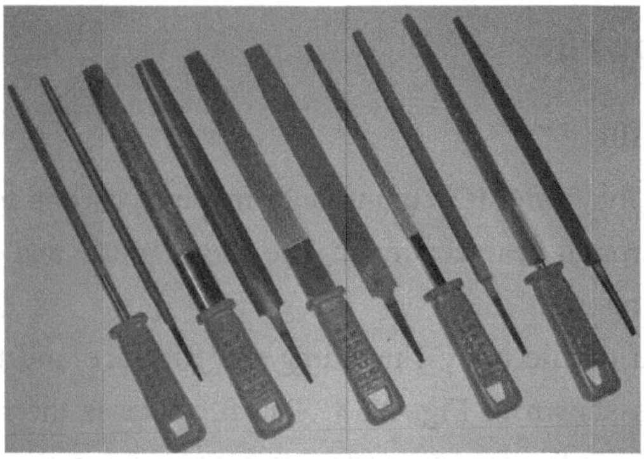

Fig. 2-20 Types of the file

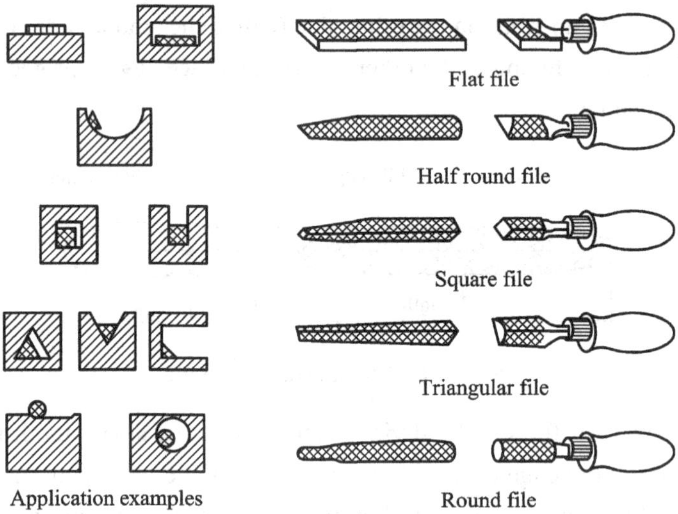

Fig. 2-21 The fitter file and their suitable processing surfaces

The shaping file are also called multi-function file or set file. They are named because they are grouped to satisfy various cross-section shapes. They are mainly used to trim small parts on the workpiece, as shown in Fig. 2-22.

Fig. 2-22 Shaping files

Files of various types and specifications can be represented by file numbers. The file serial number is composed of elements of category code-form code-specification.

Other codes of files are as follows: p stands for ordinary type; b stands for thin type; h stands for thick type; z stands for narrow type; t stands for extra narrow type; l stands for spiral type. Examples of file numbers are shown in Table 2-7.

Table 2-7 Examples of file numbers

File number	File types and specifications
Q-02-200	Pointed flat file of fitter files, with length 200 mm
Y-01-170	Flush file of irregular file type, with length 170 mm
Z-04-140	Triangular file of shaping files, with length 140 mm
Q-03h-250	Thick half round file of fitter files, with length 250 mm

(III) Maintenance method of the file

Every kind of file has its proper purpose and different application occasions. Only reasonable selection can give full play to its effectiveness and prevent premature loss of filing ability. Reasonable use and maintenance of the file can extend its service life, otherwise it will be damaged prematurely. Therefore, the following rules of use and maintenance must be paid attention to:

(1) Do not use files to file the crust or oxide skin of the workblank and hardened workpieces. Oxide and cast crust must be ground off on the grinding wheel or processed with the front end or side teeth of the file, otherwise the file will be blunt quickly.

(2) One side of the file should be used first and the other side can be used after the first side gets blunt. Because the used file surface is easy to rust. If both sides are used at the same time, the total service life will be shortened.

(3) After using the file each time, use the file brush to brush off the remaining iron filings along the iron pattern to avoid rust. During use, if iron filings are found embedded in the file pattern, it must be removed in time.

(4) No matter in use or storage, the file should not be placed overlapped or stacked with other tools and objects. It should be kept in order when storing, otherwise the file teeth could be easily damaged due to collision.

(5) The filing speed should not be too fast, otherwise the file will be easily blunt and cause slipping. The filing speed is generally about 40 times per minute, and the speed of filing hard steel parts should be slower. No pressure should be applied during the return of the filing process to avoid blunt grinding of the file teeth.

(6) Do not use a fine file as a rough file or to file soft metal (except polishing), because the file chips of soft metal are easy to be embedded in the tooth grooves of the file teeth, which makes the file slip on the surface of the workpiece.

(7) The file must not be stained with water or oil to prevent corrosion or slipping. If the file is stained with organic oil, apply chalk on the file, and then clean it with a file brush.

(8) Do not use the file as a lever to pry other objects, and do not use the file tang as an inclined iron to remove the taper sleeve to prevent damage to the file.

(9) When using small files or shaping files, do not use too much force to avoid breaking.

II. Filing method

(I) Assembly and disassembly method of the file handle

The assembly and disassembly method of the file handle are shown in Fig. 2-23.

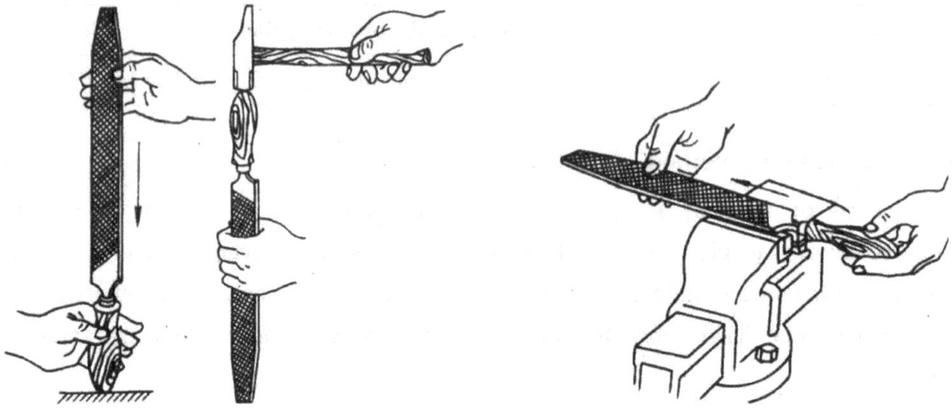

(a) Assembly method of the file handle (b) Disassembly method of the file handle

Fig. 2-23 Assembly and disassembly of the file handle

(II) Plane filing posture

Whether the filing posture is correct or not has a decisive influence on the filing quality, the application and exertion of filing force and the fatigue degree of the operator. In order to master the file posture correctly, it is necessary to pay attention to the file holding, filing posture, filing force and speed.

1. File holding

Holding the file correctly is helpful to improve the quality of filing. There are many kinds of files, so the holding methods of files must change with the size of files and the place where it is used. The holding method of plate file larger than 250 mm is shown in Fig. 2-24(a). Hold the file handle tightly with the right hand, the end of the handle is on the palm at the base of the thumb, the thumb is on the upper part of the file handle, and the rest of the fingers hold the file handle from bottom to top. The basic holding method of the left hand is to press the muscle at the base of the thumb on the file head, straighten the thumb naturally, bend the other four fingers to the palm of the hand, and hold the front end of the file with the middle finger and ring finger. There are also two ways to hold the left hand, as shown in Fig. 2-24(b). When filing, the right hand pushes the file and determines the pushing direction, and the left hand cooperates with the right hand to keep the file balanced.

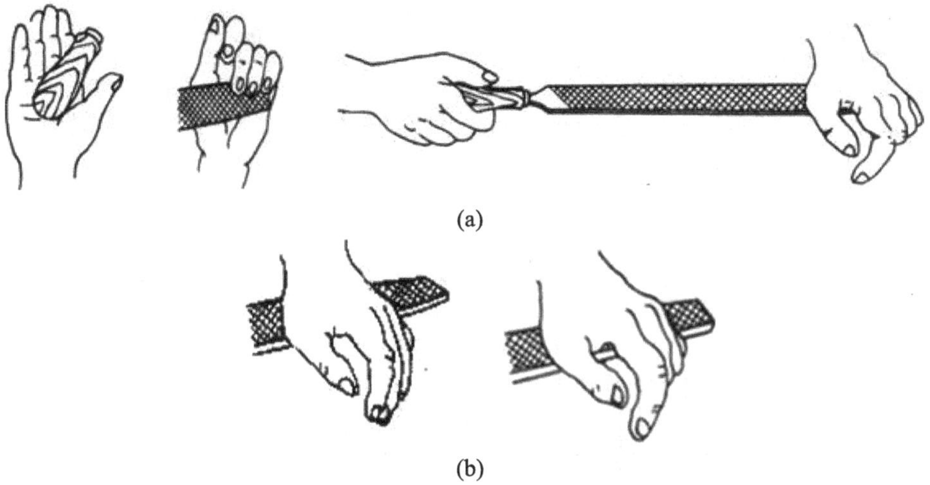

Fig. 2-24 The holding method of large plate file

2. Filing posture

Filing posture is very important for a fitter. The correct posture can improve the filing quality as well as filing efficiency and reduce labor intensity. The standing position and posture (see Fig. 2-25), filing action (see Fig. 2-26) and file pushing (see Fig. 2-27) are as follows: hold the file with both hands and place it on the workpiece, with the left arm bent, the lower arm basically parallel to the left and right direction of the workpiece filing surface, and the right lower arm basically parallel to the front and back direction of the workpiece filing surface, but naturally. When filing, the body is ahead of the file and moves forward with it, the right foot is straight and lean forward slightly, with the center of gravity on the left foot and the left knee bent. When the file reaches about 3/4 of the stroke, the body stops moving forward, and both arms continue to file the file forward to the end. At the same time, the left foot straightens naturally and moves the body's center of gravity backward with the reaction force in filing, so as to make the body return to its original position and retract the file. When the retraction of the file is nearly over, the body begins to lean forward before the file and move forward for the second time.

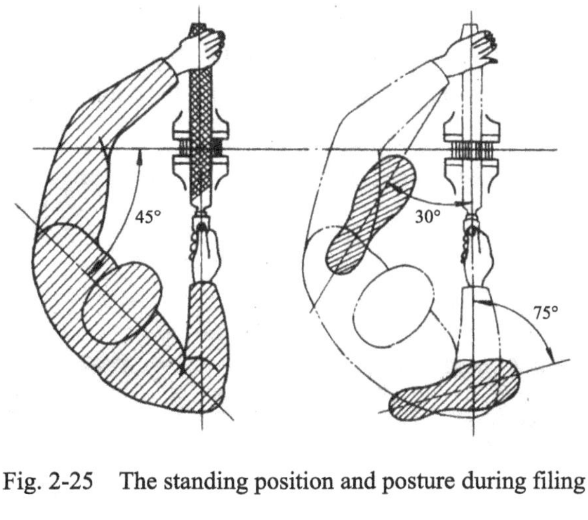

Fig. 2-25 The standing position and posture during filing

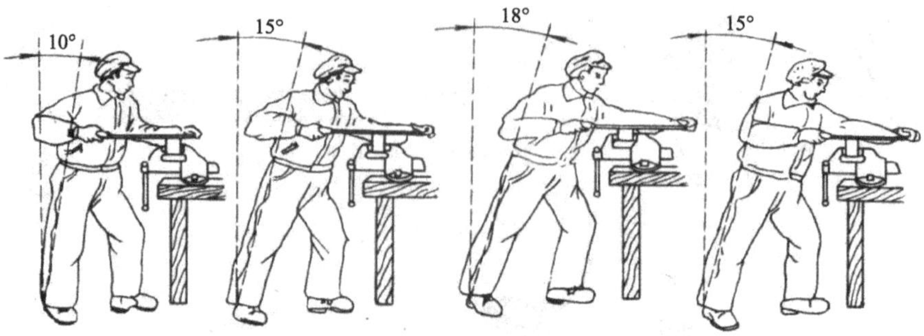

Fig. 2-26 Filing action

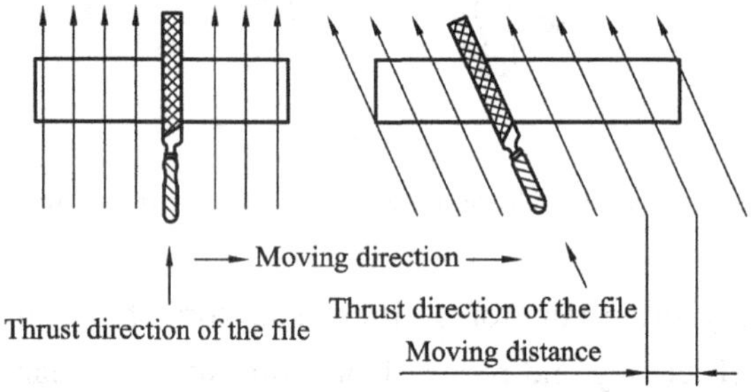

Fig. 2-27 The pushing movement of the file

3. Filing force and speed

In order to file out a flat surface, it is necessary to maintain the balance of filing force to keep the file moving in a straight line. The filing force is composed of horizontal thrust and vertical pressure. The thrust is mainly controlled by the right hand and the pressure is controlled by both hands. When filing, because the length of both ends of the file extending out of the workpiece changes at any time, the pressure of the two hands on the file must also change. At the beginning of filing, the left hand pressure should be high, the right hand pressure should be low and the thrust should be big. When filing, the pressure of the right hand should gradually increase with the file pushing, and the pressure of the left hand should gradually decrease with the file pushing, as shown in Fig. 2-28. No pressure is applied during the return stroke to reduce the wear of the teeth.

Generally, the filing speed should be about 40 times pre minute, with a slower push-out and a faster return stroke, and the movements should be naturally coordinated.

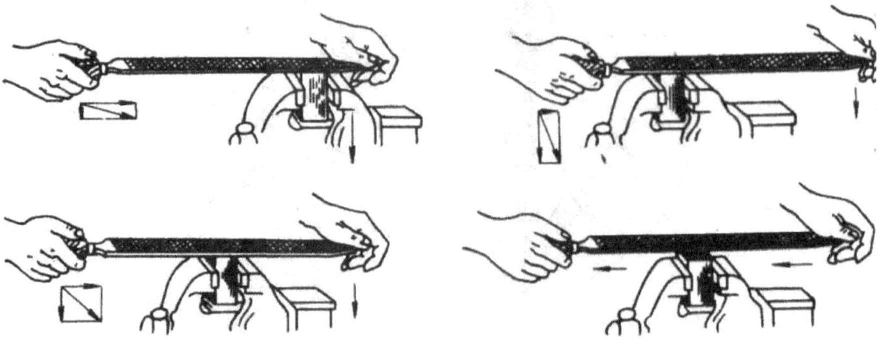

Fig. 2-28 The force by both hands when filing the plane

(III) The method of flat filing

The method of flat filing is as follows:

(1) Forward filing [see Fig. 2-29(a)]. The moving direction of the file is always consistent with the clamping direction of the workpiece. When filing a wide plane, in order to make the whole machined surface be filed evenly, it is necessary to move the file horizontally each time. This is the most basic method of filing. With this method, the straight file marks can be obtained, which are neat and beautiful. It is suitable for the final polishing filing of the workpiece surface and the small flat surfaces.

(2) Cross filing [see Fig. 2-29(b)]. The angle between the moving direction of the file and the clamping direction of the workpiece is about 30°–40°, and the file marks are crossed. Because of the large contact surface between the file and the workpiece, the file is easy to hold and stabilize. At the same time, the height of the file surface can be judged from the file marks, which is convenient for continuously correcting the filing position, but the file marks are not straight. Cross filing is generally suitable for rough filing. When there is a large file margin, the cross filing can be used first. When the margin is basically finished, the forward filing can be used to make the filing marks straight and the texture consistent.

(3) Push filing [see Fig. 2-29(c)]. Push filing is to hold the file symmetrically with both hands and push the file along the length direction of the workpiece with thumb. Push filing is suitable for filing narrow and long planes and trimming sizes.

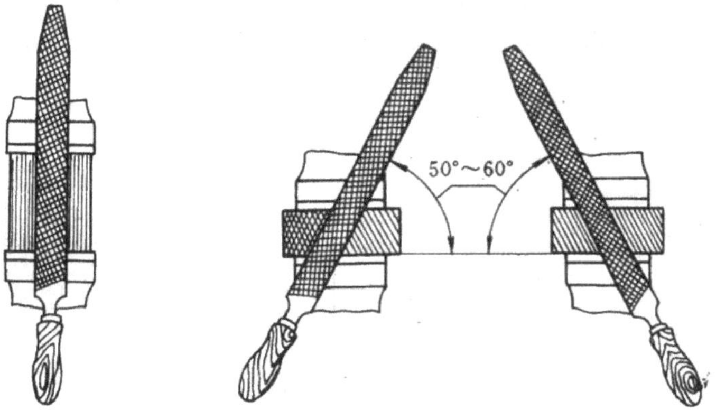

(a) Forward filing (b) Cross filing

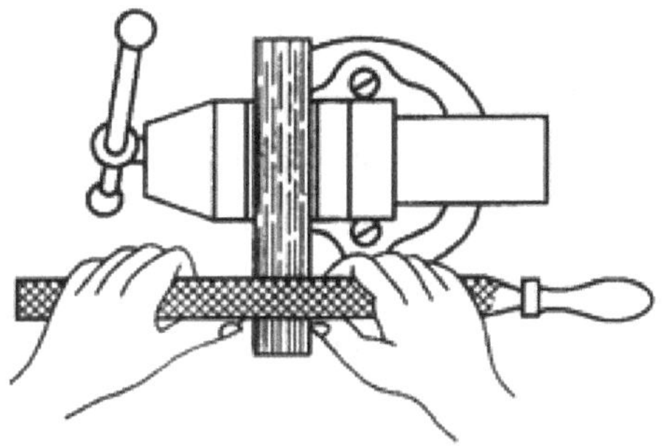

(c) Push filing

Fig. 2-29 Filing methods of the plane

(Ⅳ) Safe and civilized production knowledge of filing

(1) The file is a right-handed tool and should be placed on the right side of the bench vice. When it is placed on the bench vice, the handle of the file should not be extended outside the table, so as to avoid foot injury or file damage caused by falling.

(2) Files without a handle, a file with a cracked handle or a file without a handle hoop should not be used.

(3) When filing, the handle of the file should not hit the workpiece, so as to avoid accidents caused by the handle falling off.

(4) Do not blow the filings with your mouth or touch the surface of the file with hands.

(5) Files should not be used as crowbars or hammers.

III. Measurement of flatness, perpendicularity and parallelism

When filing a plane, a steel ruler or a knife-edge ruler is often used to check its flatness by light transmission method. The knife-edge ruler shall be inspected at several places along the longitudinal, transverse and diagonal directions of the machined surface. If the light passing through between the ruler and the plane is weak and uniform, it means that the plane is relatively straight. If the light passing through the inspection place is different, it means that the plane is uneven.

(1) Check the flatness of workpiece with a knife-edge ruler. Check the knife-edge ruler in longitudinal, transverse and diagonal directions one by one (see Fig. 2-30).

It can be inserted with a feeler gauge, and the flatness error can be determined according to the thickness of the feeler gauge (see Fig. 2-31).

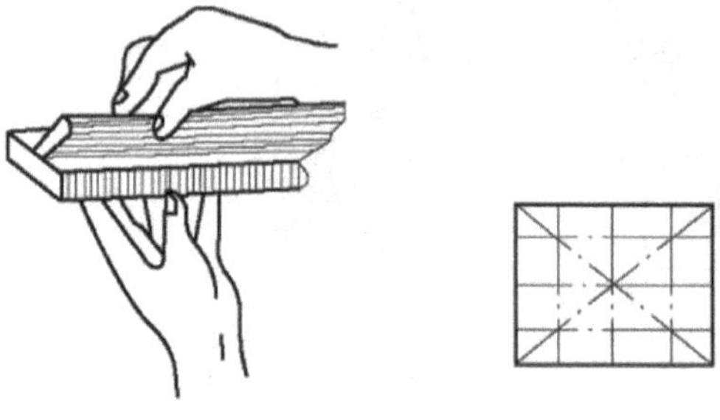

Fig. 2-30 Method of flatness detection by knife-edge ruler

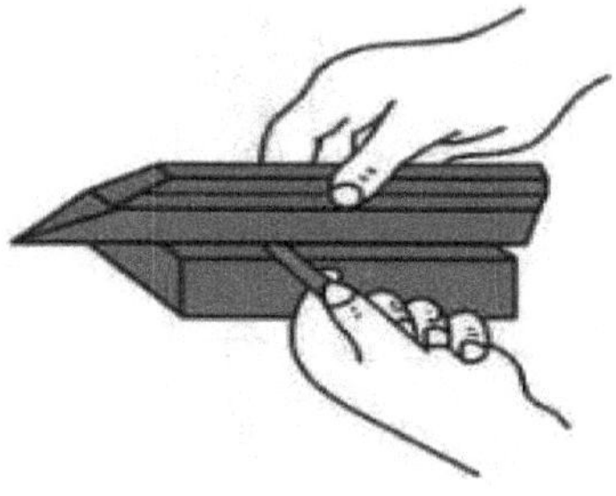

Fig. 2-31 Flatness error detection

(2) Check the perpendicularity with a square (see Fig. 2-32).

The perpendicularity is checked by the light transmission method with a square. During inspection, carefully stick the short side of the square to the surface A and move it from top to bottom until the long side is in contact with the surface B for light transmission inspection.

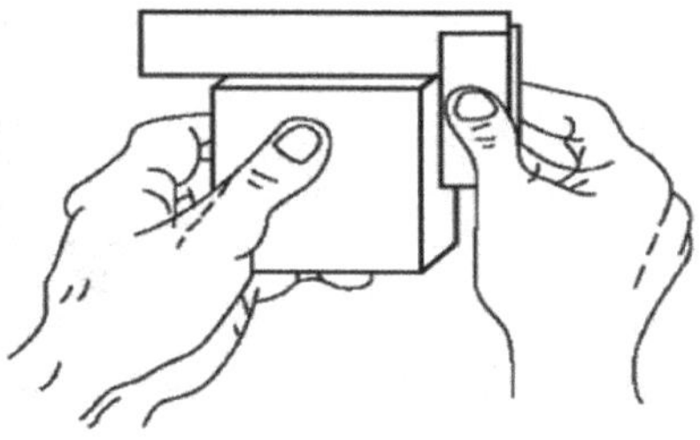

Fig. 2-32 Checking the perpendicularity

(3) When filing parallel planes, a vernier caliper

is often used to check the error between the filed surface and the reference plane, and a knife-edge ruler or a square is used to check the flatness by light transmission method.

IV. Plane filing practice

Practice filing cubes in the length direction of 45# steel (as shown in Fig. 2-33) with the size of 22 mm × 22 mm × 120 mm(as shown in Fig. 2-34).

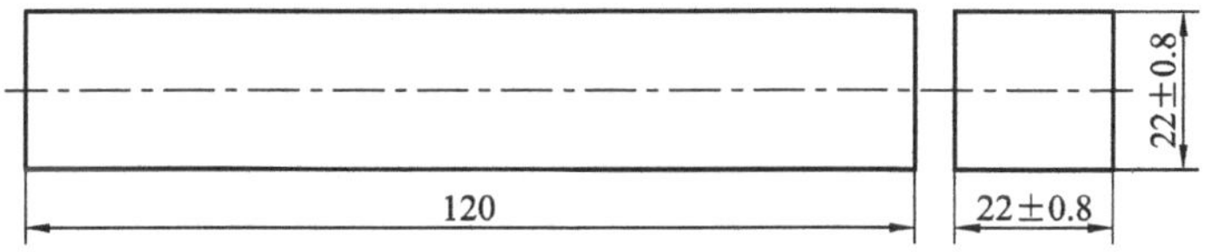

Fig. 2-33 Workblank drawing

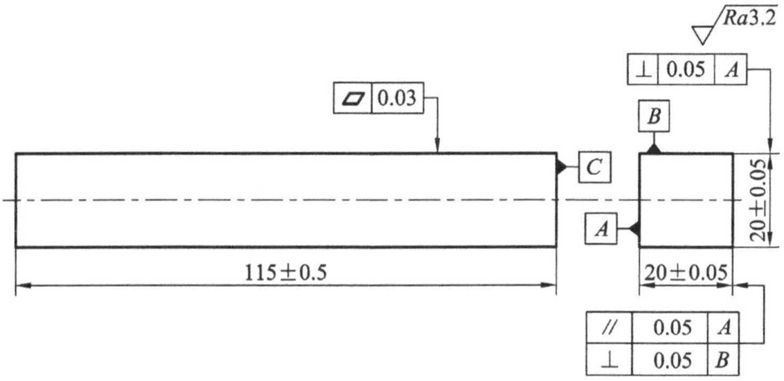

Fig. 2-34 Workpiece drawing

(I) Preparation before operation

(1) Analyze the drawings and clarify requirements: analyze and understand the parts drawing. understand the tolerance requirements of perpendicularity, parallelism and flatness.

(2) Prepare of machining tools and measuring tools: Select suitable machining tools and measuring tools according to the drawing requirements. See Table 2-8 for details.

Table 2-8 List of machining tools and measuring tools

Serial number	Name	Specification	Quantity	Remarks
1	Vernier caliper	0–150 mm	1	Precision of 0.02 mm
2	Knife-edge ruler	125 mm	1	Grade 1
3	Height vernier caliper	0–300 mm	As required	Precision of 0.02 mm
4	Square	80 mm× 50 mm	As required	0
5	Plate file	150 mm、250 mm、300 mm	1 for each	Coarse, medium and fine teeth each
6	Set file		1	
7	File brush, brush		As required	

(3) Understand the scoring standard: analyze the dimension and other technical requirements, understand the key points of score allocation, and define the important processing elements. The scoring standards are shown in Table 2-9.

Table 2-9 Scoring standard

No.	Quality inspection content	Score allocation	Scoring standard	Actual measurement	Final score
1	(20±0.05) mm (2 places)	30	No score if beyond the tolerance		
2	(115±0.05) mm	16	No score if beyond the tolerance		
3	Flatness 0.03 mm(6 places)	6×2	No score if beyond the tolerance		
4	Parallelism 0.05 mm(2 places)	2×5	No score if beyond the tolerance		
5	Perpendicularity 0.05 mm(4 places)	4×4	No score if beyond the tolerance		
6	*Ra*3.2 μm	6×1	No score if beyond the tolerance		
7	Safe and civilized production	10	No score if beyond the tolerance		

(II) Operation steps

(1) Analyze the drawing and check the workblank size.

(2) File datum plane A to meet the requirements of flatness and surface roughness.

(3) File the opposite side of datum plane A to meet the requirements of dimension (20±0.05) mm, flatness, parallelism and surface roughness.

(4) Rough and fine file datum plane B to meet the requirements of flatness, perpendicularity, surface roughness.

(5) File the opposite side of datum plane B to meet the requirements of dimension (20±0.05) mm, flatness, parallelism and surface roughness.

(6) Rough and fine file plane C to meet the requirements of flatness, perpendicularity and surface roughness.

(7) File the opposite side of datum plane C to meet the requirements of dimension (115 ± 0.05)mm, flatness, parallelism and perpendicularity.

(8) Trim burrs.

(9) Check all the parts according to the drawing and file them properly.

(Ⅲ) Relevant precautions

(1) Filing is an important basic operation of the fitter. Correct posture is the basis of mastering filing skills, so it is necessary to master it.

(2) In the first practice, there will be various incorrect postures, especially the uncoordinated movements of the body and hands. It is necessary to correct them in time. If the incorrect posture becomes a habit, it will be difficult to correct.

(3) When practicing postures, pay attention to the force of both hands to keep the file moving in a straight line on the workpiece.

(4) One sidedness should be avoided in processing. The requirements of dimension and angle accuracy should not be affected in order to reach the plane accuracy; the flatness and parallelism should be ignored because of the angle; or ignore other requirements to reduce the surface roughness. In a word, all accuracy requirements should be taken into account during processing.

Task Four Hole machining

※ Knowledge objectives and skill objectives

(1) Understand the structure and geometric parameters of twist drills, countersink drills and reamers.

(2) Understand the structure, specification, performance and use of bench drills and vertical drills.

(3) Be familiar with several basic clamping methods of the workpiece and the selection method of rotation speed during drilling.

(4) Master the method of marking and drilling, and be able to drill general holes.

(5) Master the methods and precautions of countersinking.

(6) Master the knowledge and precautions of reaming method, reaming amount and reaming speed.

I. Drilling

(I) Overview of drilling

Drilling is a method of machining a hole in a solid material with a drill. When drilling with a drilling machine (see Fig. 2-35), the workpiece is installed and clamped on the drilling machine table and fixed, and the drill bit is installed on the main shaft of the drilling machine (or on the drill

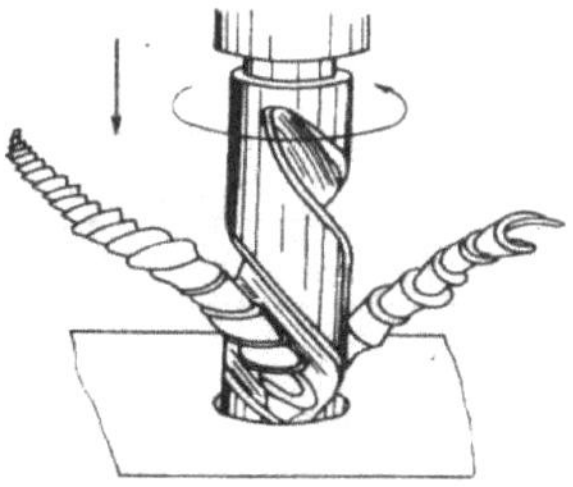

Fig. 2-35 Hole drilling

chuck connected with the main shaft). The drill rotates(main motion) and moves downward along the axis of the drill bit (feeding motion).

When drilling, due to the poor rigidity and accuracy of the drill bit, the machining accuracy is not high, generally IT10–IT11, and the roughness of surface is⩾12.5 μm. Therefore, drilling can only be used to process holes with low accuracy or be used as rough machining of holes.

(II) Structure of the twist drill

There are many types of drill bits, such as twist drills, flat drills, deep hole drills, center drills, etc. The twist drill is the the most commonly used type of drill. It is generally made of high-speed steel, and its structure is composed of shank, neck and working part, as shown in Fig. 2-36.

The shank of the drill bit is the connecting part with the drilling machine, which is used to transmit the required torque and axial force when drilling. There are two types of twist drills: straight shank type and taper shank type. Generally, the drill bit with diameter less than 13 mm is made into straight shank, and the drill bit with diameter more than 13 mm is made into taper shank.

The neck is used as a groove when grinding twist drill. The specifications, materials and trademarks of the drill are printed on the neck.

The working part is composed of cutting part and guiding part. The cutting part consists of five cutting edges (two main cutting edges, two auxiliary cutting edges and one chisel edge) and six faces (two rake faces, two flank faces and two auxiliary flank faces). The cutting part mainly plays the role of cutting the workpiece. The function of the guiding part in the cutting process is not only to ensure the correct direction of the drill bit, and to polish the hole wall, but also works as the backup of the cutting part.

The solid part of the working part of the drill bit along the axis line is drill core. It is used to connect two blades to maintain the strength and rigidity of the drill. The drill core gradually increases from the cutting part to the shank, forming a vertebral body.

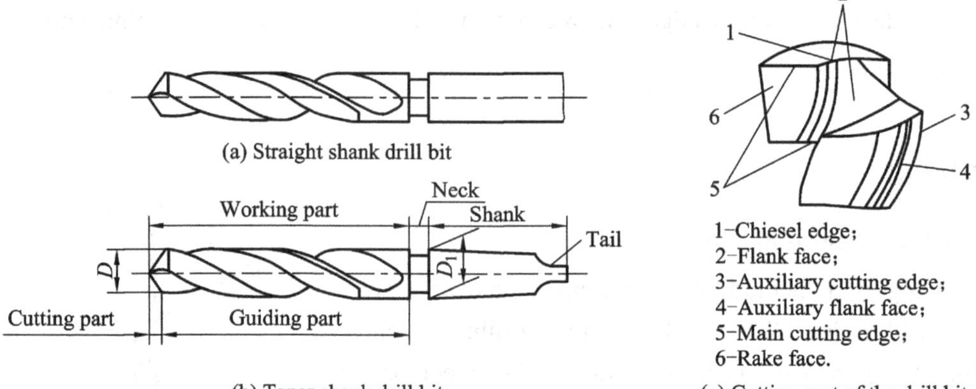

(a) Straight shank drill bit

(b) Taper shank drill bit

(c) Cutting part of the drill bit

Fig. 2-36 Structure of the twist drill

(III) Drilling machine

There are many types and forms of drilling machines, but apart from multi-head drilling machines and specialized drilling machines, there are three types of drilling machines commonly used in daily drilling: bench drilling machines, vertical drilling machine and radial drilling machines, as shown in Fig. 2-37.

1. Bench drilling machine

The speed change of the bench drilling machine is realized by a set of pulleys installed on the motor main shaft and the drilling machine. There are five different speeds, and the machine should stop when changing speed.

The bench drilling machine has the advantages of high speed, high efficiency, convenient and flexible use, and is suitable for drilling small workpieces. However, the bench drilling machine is not suitable for the processing of countersinking and reaming , because its minimum speed is high.

(a) Bench drilling machine (b) Vertical drilling machine (c) Radial drilling machine

Fig. 2-37 Drilling machines

2. Vertical drilling machine

Vertical drilling machine is a common type of drilling machine. It is mainly composed of main spindle, main spindle gearbox, feed box, column, workbench and base. Its specification is expressed by the maximum drilling diameter, such as 25, 35, 40, 50, etc. By operating the handle, the feed gearbox can move up and down along the column guide rail to adjust the distance between the main spindle and the workbench. Shaking the handle of the workbench can also move the workbench up and down along the column guide rail to adapt to the processing of workpieces of different sizes.

3. Radial drilling machine

When drilling holes on large workpiece or drilling lots of holes on the same workpiece, a radial drilling machine can be used. Radial drilling machine is to move the drill shaft to align with the drilling center for drilling, so the operation is labor-saving and flexible. Its structure is mainly composed of base, workbench, column, spindle gearbox and radial arm.

When drilling, according to the needs of workpiece processing, the radial arm can move up and down along the column and rotate 360° around the column. The spindle gearbox can move in a large range along the radial arm guide rail, which is convenient for locating the position between the drill bit and the drilling hole.

(IV) Key points of drilling operation and maintenance of drilling machine

When drilling on the drilling machine, the workpiece is fixed, and the drill bit rotates and moves axially to complete the deep drilling.

1. Key points of drilling operation

(1) Before drilling, wipe the surface sliding part of the drilling machine clean, fill in lubricating oil, and move the operating handle to the correct position, and then start at a low speed, after a few minutes of trial operation, and wait until the mechanical transmission and lubrication system operate normally, then start drilling.

(2) During the clamping process, the workpiece should be carefully calibrated to ensure that the center line of the drilling hole is perpendicular to the worktable of the drilling machine. When setting the tool, observe the alignment of the cross edge of the drill bit to the punching hole of the sample from different directions.Before drilling, spot a shallow hole, and then drill the hole formally after confirming that it is correct.

(3) Before the drill bit is clamped, the shank and the taper hole of the drilling machine spindle should be wiped clean. After the drill bit is installed, slowly rotate the main spindle of the drilling machine to check whether the drill bit is upright. If it swings, it can be clamped in different directions to adjust the swinging to the minimum value.

(4) At the beginning of drilling, the drill bit should contact the workpiece slowly. Do not hit the workpiece with the drill bit to avoid damaging the drill bit. When drilling holes in the rough surface of the workpiece, manual feed should be used at the beginning. In this way, when encountering hard point, the drill bit can retreat to avoid breaking the cutting edge.

(5) When the speed needs to be changed, the machine must be stopped first to avoid damaging the gear or other parts. When moving the speed shifting handle, it should be placed in the specified position. If the handle is found to be malfunctioning or cannot be moved to the required position, it should be checked and adjusted, and it should not be moved with brute force.

(6) When the operator leaves the drilling machine or changes tools or workpieces, or when the main power supply is suddenly cut off, the switch of the drilling machine should be turned off.

2. Maintenance of drilling machine

When using machine tools, in order to reduce equipment accidents, extend the service life of machine tools, and improve equipment integrity, in addition to reasonable use of machine tools in accordance with operating regulations, the maintenance of machine tools must be done carefully. Before using the drilling machine, the operator should carefully check and wipe all parts of the drilling machine and oil it for maintenance to keep the drilling machine lubricated and clean. In case of any accident, the operator should eliminate it in time and make a record.

(V) The method of marking and drilling

1. Marking of the workpiece during drilling

According to the position and size requirements of the drilling hole, draw the cross center line of the hole, punch the center sample (the punching point shall be small and the position shall be accurate), and draw the circle line of the hole according to the size of the hole. For holes with larger diameter, several inspection circles of different sizes should be drawn [see Fig. 2-38(a)], so as to check and correct the drilling position when drilling. When the position and size of the drilling hole are strictly required, in order to avoid the deviation caused by striking the center punch, several boxes of different sizes [see Fig. 2-38(b)] symmetrical to the center line of the hole can be directly drawn as the inspection line when drilling, and then the center sample punching hole can be enlarged to make the hole accurate.

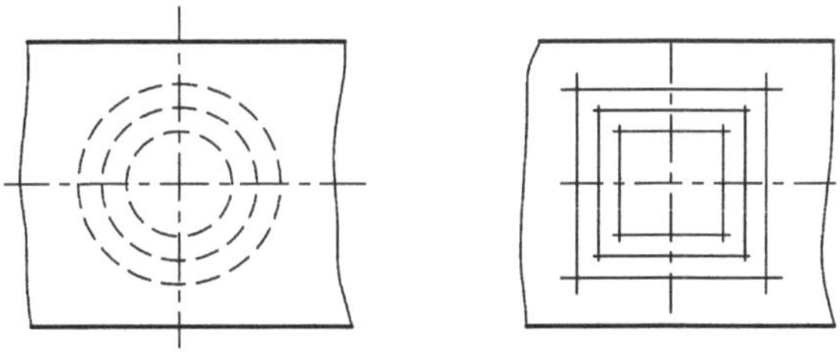

(a) Checking circle (b) Checking square

Fig. 2-38 Checking form of hole position

2. Workpiece clamping

When drilling a workpiece, different clamping (positioning and clamping) methods should be adopted according to the different shape of the workpiece and the force of drilling (or the diameter of the hole) to ensure the quality and safety of the drilling. The commonly used basic clamping methods are as follows:

(1) Flat workpiece can be clamped with flat tongs[see Fig. 2-39(a)]. When clamping, the workpiece surface should be perpendicular to the drill bit. When drilling a hole with a diameter bigger than 8 mm, the flat tongs must be fixed with bolts and pressing plates. When drilling through holes with the workpiece clamped by the bench vice, the bottom of the workpiece should be padded with an iron gasket, so as to free up the falling part of the drill, to avoid damaging the bench vice.

(2) Cylindrical workpiece can be clamped with V-shaped iron [see Fig. 2-39(b)]. During clamping, the axis line of the drill should be perpendicular to the symmetrical plane of the V-shape to ensure that the center line of the drilled hole passes through the axis line of the workpiece.

(3) For larger workpiece with a diameter of more than 10 mm, the method of pressing plate clamping can be used for drilling [see Fig. 2-39(c)].

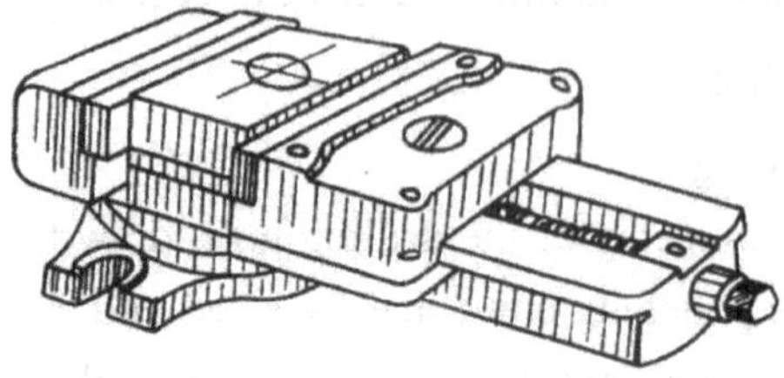

(a) With flat tongs

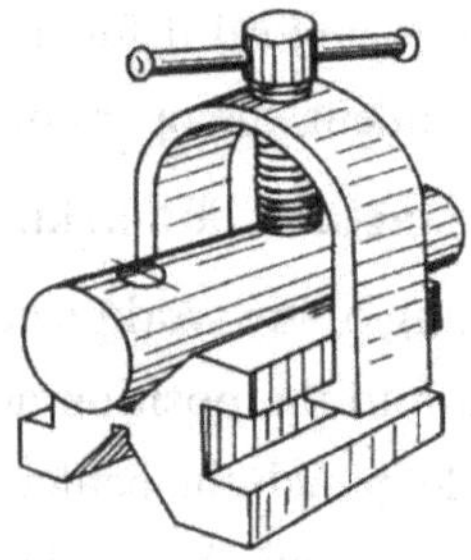

(b) With V-shaped iron

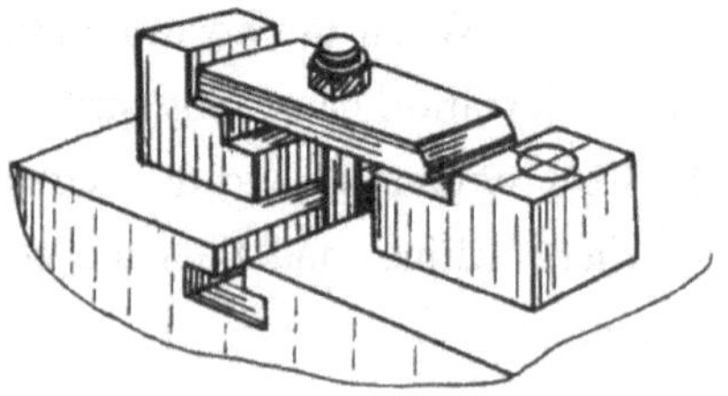

(c) With pressing plate

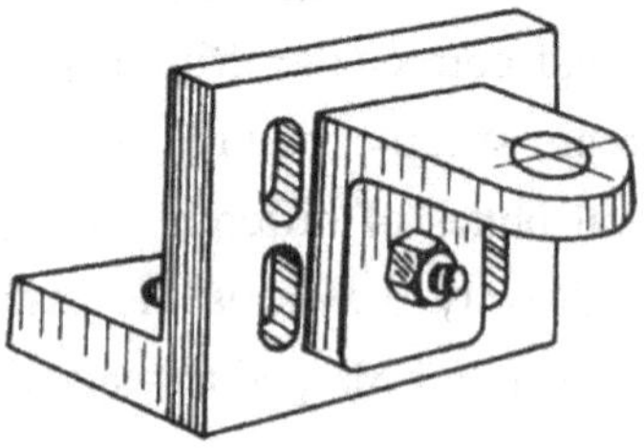

(d) With angle iron

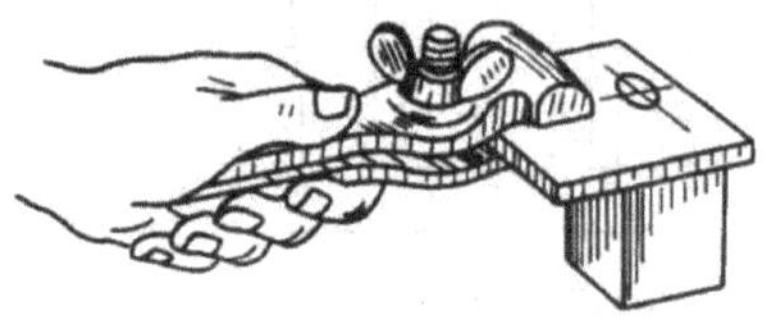

(e) With hand vice

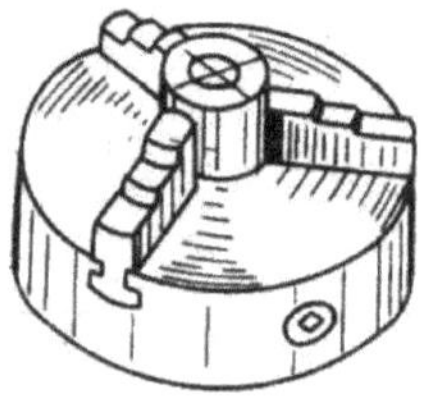

(f) With three-jaw self-centering chuck

Fig. 2-39 Workpiece clamping methods

(4) If the bottom surface is uneven or the machining datum is on the side, the angle iron can be used for clamping [see Fig. 2-39(d)]. Because the axial drilling force acts outside the installation plane of the angle iron, so the angle iron must be fixed on the drilling machine workbench with a pressing plate.

(5) When drilling a small hole on a small workpiece or thin plate, the workpiece can be placed on the positioning block and clamped with a hand vice [see Fig. 2-39(e)].

(6) When the end face of cylindrical workpiece is drilled, the three-jaw self-centering chuck can be used for clamping [see Fig. 2-39(f)].

3. Assembly and disassembly of drill

(1) Assembly and disassembly of the straight shank drill bit. The straight shank drill bit is clamped by the drill chuck. First, insert the drill shank into the three jaws of the drill chuck, and the clamping length shall not be less than 15 mm. Then, use the drill chuck key to rotate the outer sleeve, so that the ring nut can drive the three jaws to move, acting clamping or loosening (see Fig. 2-40)

(2) Assembly and disassembly of taper shank drill bit. For the taper shank drill bit, the Morse taper of the shank is directly connected with the spindle of the drilling machine. When connecting, the taper shank of the drill and the hole of the spindle must be wiped clean, and the

longer direction of the rectangular tang should be consistent with the center line of the waist hole in the spindle. Accelerated impulse should be used for one-time assembly [see Fig. 2-41(a)]. When the taper shank drill bit is smaller than the taper hole of the spindle, a transition sleeve [see Fig. 2-41(b)] can be added to connect it. To disassemble the drill bit in the sleeve and the drill bit on the spindle of the drilling machine, the inclined iron is used to knock into sleeve or the waist shaped hole in the spindle of the drilling machine, the side of the inclined iron with arc should be placed upward, together with the downward force of the inclined iron slope, the drill bit and the sleeve or the spindle can be separated [see Fig. 2-41(c)]. The drill bit should be installed firmly on the spindle of the drilling machine, and the radial runout should be the minimum when rotating.

Fig. 2-40 Clamping with drill chuck

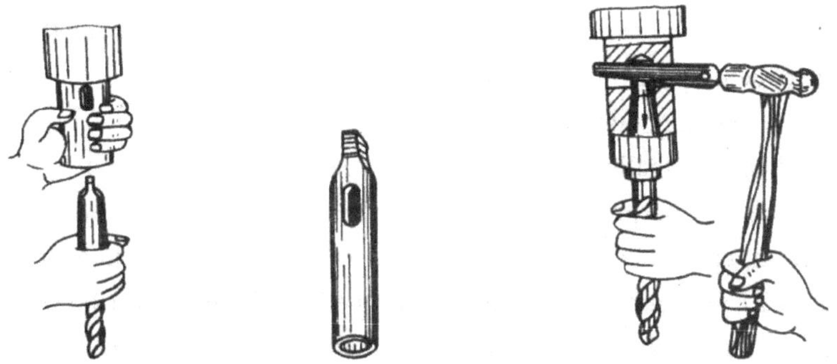

(a) (b) (c)

Fig. 2-41 Assembly and disassembly of taper shank drill bit and the transition taper sleeve

4. Selection of the drilling speed

When selecting the drilling speed, the allowable cutting speed v of the drill bit must be determined first. When drilling iron castings with high speed steel bits, $v = 14-22$ m/min; when drilling steel parts, $v = 16-24$ m/min; when drilling bronze or brass, $v = 30-60$ m/min; when the hardness and strength of the workpiece material are high, the speed should be relatively small; when the drill diameter is small, the speed is accordingly small (take $\phi 16$ mm as the median value); when the drilling depth $L>3d$, the value should be multiplied by a correction factor of 0.7–0.8. Then use the following formula to calculate the drilling machine speed n:

$$n = \frac{1000v}{\pi d}$$

In the formula, v represents cutting speed (unit in m/min), d represents drill diameter(unit in mm), and the unit of n is r/min.

5. Cutting amount

Reasonable selection of cutting amount during drilling can improve drilling accuracy and production efficiency, and can prevent machine overload or drill damage.

Since the amount of back engagement of the cutting edge is determined by the diameter of the drill bit, the cutting amount during drilling is determined by the drilling speed and the feeding rate. A higher drilling speed v or feeding rate f can improve production efficiency. But too high drilling speed will cause strong friction and shorten the service life of the drill bit. If the feeding rate is too large, although it has little effect on the service life of the drill bit, it will directly affect the residual area of the machined surface, and the larger the residual area is, the rougher the machined surface is.

6. Tripping out

When drilling, first align the drill bit to the center of the hole to drill a shallow hole, observe whether the drilling position is correct, and constantly correct it to make the shallow hole coaxial with the marking circle. Space borrowing method: If the deviation is small, the workpiece can be forced to move in the opposite direction of the deviation while drilling to achieve a gradual correction; if the deviation is relatively big, several holes can be punched or grooves can be carved with oil groove chisel in the correction direction to reduce the drilling resistance here and achieve the purpose of correction.

7. Feeding operation by hand

When the shallow hole drilled at the beginning meets the requirement of position, the workpiece can be pressed to complete the drilling. When feeding by hand, the feeding force should not cause the drill bit to bend, so as to avoid the drilling axis skew. When drilling holes with small diameter or deep holes, the feeding force should be small, and the drill should be retracted frequently to avoid chip jam or breaking of the drill bit. Generally, when the drilling depth reaches 3 times of the diameter, the drill must be retracted to remove chips; when the drilling is about to pass through, the feeding force must be reduced to prevent sudden excessive feeding and increase the cutting resistance, causing the drill bit to break, or even accidents caused by the rotation of the workpiece.

8. Cutting fluid

In order to cool the drill bit, reduce the friction between the drill bit and the workpiece and chips during drilling, and eliminate the built-up edge on the surface of the drill bit and the workpiece, thereby reducing the cutting resistance, increasing the service life of the drill bit and improving the quality of the surface of the processed hole, enough cutting fluid must be added when drilling. When drilling steel parts, 3%–5% emulsion can be used for continuous filling; when drilling cast iron, 5%–8% emulsion can be used for continuous filling.

(VI) Safety knowledge during drilling

(1) Do not wear gloves when operating the drilling machine, and the cuffs must be tied tightly. Girls must roll their long hair into the work cap.

(2) the workpiece must be clamped carefully, especially when drilling a diameter hole in a small workpiece, the clamping must be firm. When the hole is about to be drilled through, the feeding force should be minimized.

(3) Before starting the drilling machine, check whether there is a drill chuck key or inclined iron inserted on the spindle.

(4) When drilling, do not use hands or cotton yarn heads or the mouth blowing to remove

the chips. The brush must be used to remove them. When drilling long strips of chips, use a hook to break them and then remove them.

(5) The head of the operator is not allowed to be too close to the rotating spindle. When stopping, the spindle should stop naturally. It is not allowed to use hand or reverse rotation to stop it.

(6) It is strictly forbidden to assemble or disassemble the workpiece, inspect the workpiece and change the spindle speed while drilling. The operations above must be carried out in the stopped state.

(7) When cleaning the drilling machine or adding lubricating oil, the power supply must be cut off.

II. Countersinking

Countersinking is a method of machining a flat-bottomed or tapered counterbore with a countersink (or a modified drill bit). The application of countersinking are shown in Fig. 2-42.

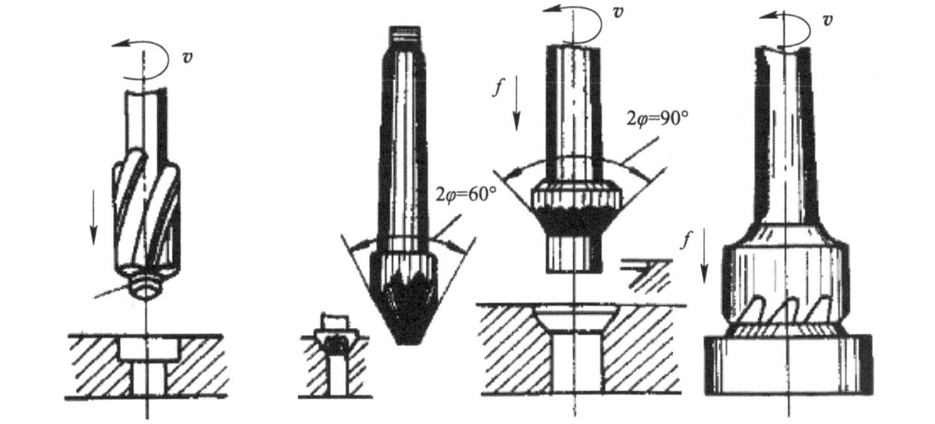

(a) Countersinking cylindrical hole (b) Countersinking tapered hole (c) Countersinking face on the hole

Fig. 2-42 Application of countersinking

(I) Types and characteristics of countersink

The types of countersink are generally divided into three types: cylindrical countersink, tapered countersink and face countersink.

(1) Cylindrical countersink. The drill used to countersink cylindrical hole is cylindrical countersink, as shown in Fig. 2-42(a).

(2) Tapered countersink. The drill used to countersink tapered hole is tapered countersink, as shown in Fig. 2-42(b).

(3) Face countersink. The drill used to countersink face on the hole is face countersink, as shown in Fig. 2-42(c).

(II) Methods of countersinking

The countersinking method is basically the same as the drilling method, but the cutting tool is easy to vibrate when reaming, especially the countersink drill which is transformed from the twist drill. It is easy to produce vibration marks on the countersink face or tapered surface, which affects the quality of countersinking. When the twist drill is transformed into the countersink drill, the clearance angle and the rake angle at the outer edge should be appropriately reduced to prevent tool sticking. The two cutting edges should be symmetrical to maintain cutting balance. Try to use shorter drill bit to reduce vibration.

(III) Precautions of countersinking

(1) Try to use shorter drill bit to transform into the countersink drill, and the height of the two cutting edges should be consistent and the angle should be symmetrical when grinding. At the same time, the drill bit should be polished with oilstone after being ground on the grinding wheel, so as to make the cutting uniform and stable, and reduce the vibration during processing.

(2) The cutting speed during countersinking should be lower than that of the drilling, generally 1/2–1/3 of the drilling cutting speed. At the same time, the axial resistance of countersinking is small, so the feeding pressure by hand should not be too large and should be uniform.

(3) In order to control the depth of the countersinking, the feeding depth of the drilling machine spindle can be adjusted and positioned with the depth gauge or positioning nut on the drilling machine before the countersinking.

(4) Safe and civilized operation should be ensured.

III. Reaming

Reaming is a method of removing the trace metal layer from the hole wall of the workpiece with a reamer to improve its dimensional accuracy and reduce its surface roughness. Due to the large number of teeth and small cutting amount of reamer, the cutting resistance of reaming is small, the guiding property is good, and the machining accuracy is high. Generally, The accuracy can reach IT9–IT7, and the surface roughness can reach $Ra1.6$ μm.

(I) Structure and types of reamer

There are many kinds of reamer, which can be divided into hand reamer and machine reamer according to their usage methods. According to the shape of the reamer, it can be divided into cylindrical reamer and conical reamer. According to the structure of reamer, it can be divided into integral reamer and adjustable reamer. The following types are commonly used by fitters.

1. Integral cylindrical reamer

There are two kinds of integral cylindrical reamers: machine reamer and hand reamer. Their structures are shown in Fig. 2-43. The reamer is composed of machine part, neck and shank. The machine part includes cutting part and calibration part. There are main structural parameters: diameter (D), cutting cone angle (2φ), rake angle (γ_0) and clearance angle (α_0) of the cutting part, the land width (f) of calibration part, number of teeth (z) and chip holding groove of the calibration part, etc.

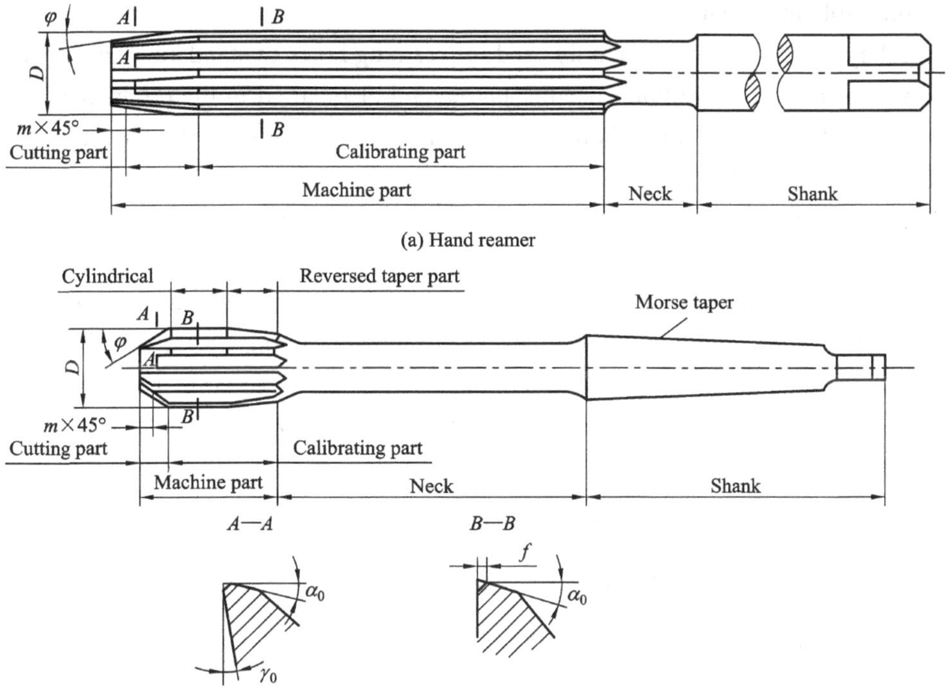

(a) Hand reamer

(b) Machine reamer

Fig. 2-43 Integral cylindrical reamer

Generally, the teeth of machine reamer are evenly distributed on the circumference, as shown in Fig. 2-44(a). When machine reamer reams holes, the reamer is driven by the machine tool to rotate continuously, and some accuracy is guaranteed by the machine tool, so there is a certain difference between machine reamer and hand reamer in structure. In order to obtain higher reaming quality, the teeth pitch of hand reamer is unevenly distributed on the circumference, as shown in Fig. 2-44(b). However, for the convenience of manufacture and measurement, the reamer with unequal pitch is often made into unequal pitch that is 180° symmetrical. Using a reamer with unequal pitch, the cutting edge will not stop at the same place when reaming, which will cause dents on the hole wall, so that the hard points can be removed and the reaming quality can be improved.

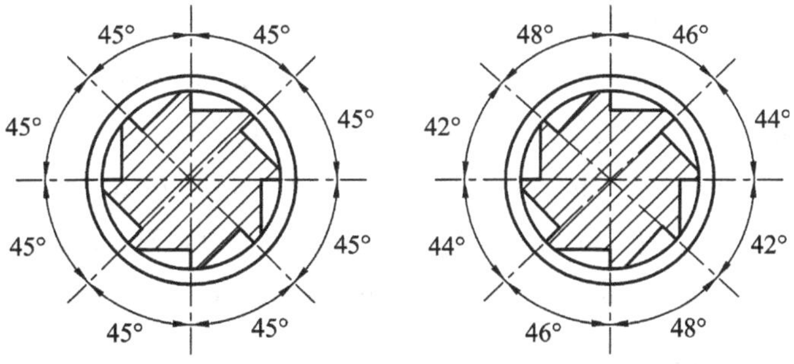

(a) Even distribution (a) Uneven distribution

Fig. 2-44 Distribution of reamer teeth

2. Adjustable hand reamer

Integral cylindrical reamer is mainly used for reaming holes of standard diameter series. If a small number of non-standard holes need to be reamed or repaired in the production of a single piece, an adjustable hand reamer should be used, as shown in Fig. 2-45. The adjustable hand reamer is composed of blade body, blade and adjusting nut. There are six straight grooves with inclined bottom on the cutter body. this blades with the same slope are embedded in the grooves, and are compressed with nuts at both ends. Adjusting nuts at both ends can make the blade move along the inclined slot in the groove, thus changing the diameter of the reamer. The diameter of standard adjustable reamer ranges from 6 mm to 54 mm. The adjustable hand reamer body is made of 45# steel, the blade with a diameter less than or equal to 12.75 mm is made of alloy tool steel, and the blade with a diameter greater than 12.75 mm is made of high speed steel.

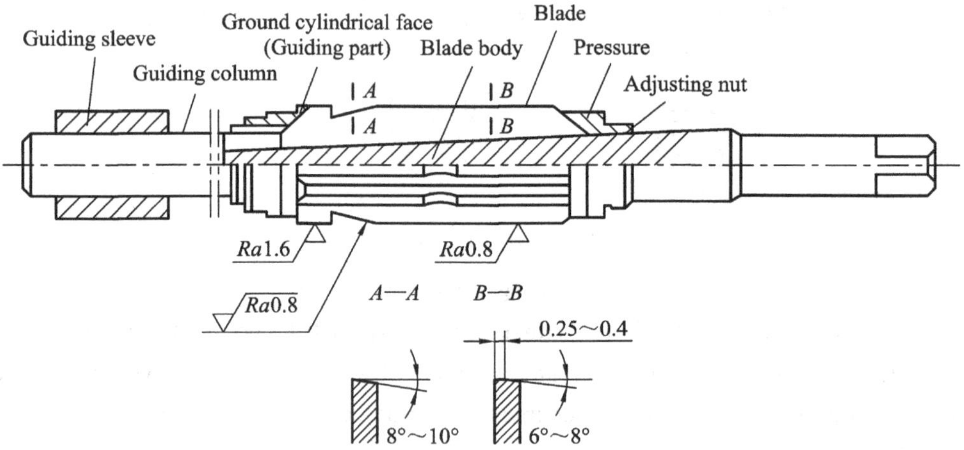

Fig. 2-45 Adjustable hand reamer

3. Taper reamer

Taper reamer is used for reaming taper holes. The commonly used types are shown as follows:

(1) The 1:50 taper reamer is used to ream fixed pin holes.

(2) The 1:10 taper reamer is used to ream the taper hole on the couplings.

(3) The Morse taper reamer is used to ream Morse taper holes.

(4) The 1:30 taper reamer is used to ream taper holes on sleeve cutters.

The tapering degree of 1: 10 taper hole and Morse taper hole is relatively large. In order to make the reaming less laborious, this kind of reamer is generally made into a set of 2 to 3 pieces, one of which is a fine reamer, and the rest are coarse reamers. The cutting edge of the coarse reamer is provided with spiral chip separation grooves to reduce the cutting load.

(II) Reaming method

1. Determination of reaming allowance

Reaming allowance is the allowance in the diameter direction left by the previous process (drilling or reaming). Whether the reaming allowance is appropriate or not has a great influence on the surface roughness and accuracy of the reamed hole. The reaming allowance should not be too large or too small. When the reaming allowance is too large, it will increase the load of the cutting edge, increase the deformation, and accordingly reduce the accuracy of the reamed hole size and increase the surface roughness; when the reaming allowance is too small, it is difficult to correct the residual deformation of the previous process, and the original cutting trace can not be removed, which affects the shape accuracy and surface roughness of the hole. When reaming with a high-speed steel standard reamer, the reaming allowance can be referred to as shown in Table 2-10.

Table 2-10 Reaming allowance

Reaming diameter/mm	<5	5–20	21–32	33–50	51–70
Reaming allowance/mm	0.1–0.2	0.2–0.3	0.3	0.5	0.8

2. Cutting speed(v) and feeding rate(f) of reaming

In order to obtain smaller surface roughness during machine reaming, it is necessary to avoid chip accumulation, reduce cutting heat and deformation, and so the smaller cutting speed should be selected. When reaming steel parts with high-speed steel reamer, v = 4–8 m/min; when reaming iron castings, v = 6–8 m/min; when reaming copper parts, v = 8–12 m/min; when reaming steel parts and iron castings, f = 0.5–1 mm/r; when reaming copper or aluminum materials, f = 1–1.2 mm/r.

3. Cutting fluid

The reamed chips are usually very fine and easy to adhere to the cutting edge, or even sandwiched between the hole wall and the edge of the calibration part to roughen the machined surface. In the reaming process, too much heat accumulation will also cause the deformation or enlargement of the hole diameter of the workpiece and reamer. Therefore, appropriate cutting fluid must be used to reduce friction and heat dissipation, and at the same time to flush out the chips in time.

4. Key points of reaming

(1) The workpiece should be clamped properly to prevent deformation of the workpiece, so as to avoid springback of the deformed part after reaming, which will affect the accuracy of the hole.

(2) During hand reaming, the force of both hands should be balanced to keep the stability of reaming to avoid the flaring of the hole and the enlargement of hole diameter caused by the swing of reamer.

(3) As the reamer rotates, press gently with both hands to make the reamer feed evenly. At the same time, the stop position of the reamer is constantly changed to prevent vibration marks

caused by continuous stop at the same position.

(4) When reaming, pay attention to whether the coaxiality error of machine tool spindle, reamer and workpiece hole meet the requirements.

(5) After reaming, the reamer should be stopped after exiting the hole, otherwise the hole wall will have blade marks and the hole will also be roughened when exiting.

(6) In the process of reaming, the cutting fluid should be selected reasonably according to the requirements of workpiece material and reaming accuracy.

(III) Precautions for reaming

(1) Reamer is a tool of fine machining. The cutting edge must be protected to avoid collision. If there is burr or chip adhesion on the cutting edge, it can be carefully polished with oil stone.

(2) The chip removal function of the reamer is poor, and the chips must be removed frequently to prevent the reamer from being jammed.

(3) When reaming the fixed taper pin hole, because of the small taper degree and self-locking, the feeding amount should not be too large to prevent the reamer from being jammed or even broken.

(4) Be familiar with the common problems and causes in reaming, so as to pay special attention to them in practice.

IV. Hole machining practice

The drawing of hole machining practice is shown in Fig. 2-46.

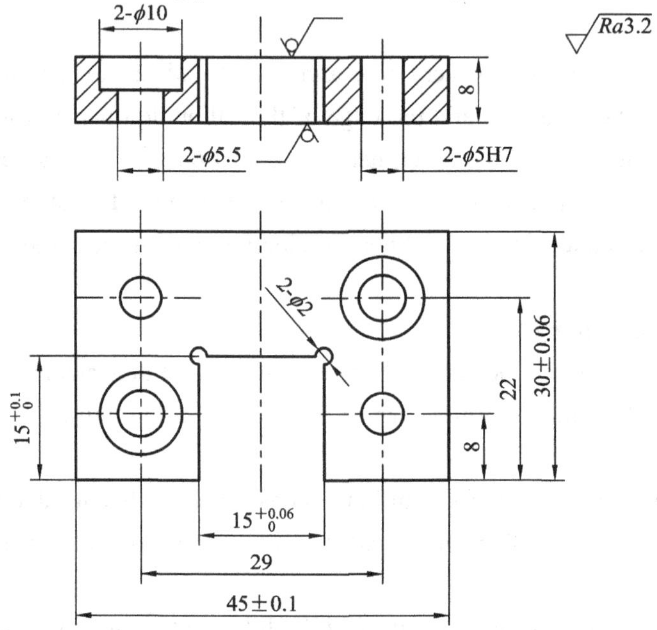

Fig. 2-46 Right guide plate

(I) Preparation before operation

(1) Analyze the drawings and clarify the requirements: formulate processing technology and make preparation before processing; determine the tools and measuring tools as required.

(2) Select machining tools and measuring tools: according to the requirements of the drawings, choose matching tools and measuring tools reasonably. See Table 2-11 for details.

Table 2-11 List of machining tools and measuring tools

No.	Name	Specification	Quantity	Remarks
1	Vernier caliper	0–150 mm	1	Precision of 0.02 mm
2	Knife-edge ruler	125 mm	1	Grade 1
3	Height vernier caliper	0–300 mm	As required	Precision of 0.02 mm
4	Micrometer	0–25 mm	1	Precision of 0.02 mm
5	Square			
6	Straight shank twist drill	ϕ2 mm、ϕ4.8 mm、ϕ5.5 mm	1 for each	
7	Reamer	ϕ5 mm	1 for each	
8	Countersink	ϕ10 mm	1 for each	
9	Saw bow, Saw blade, Hand hammer, Scribing compass, Marking pin, Sample punch, Steel ruler, Soft jaws, File brush, Brush, Motor oil, Pencil, etc.		As required	

(3) Understand the scoring standard: analyze the dimension and its technical requirements, understand the key points of distribution, and clarify the important processing elements. The scoring standard is shown in Table 2-12.

Table 2-12 Scoring standard

No.	Inspection items	Score allocation	Scoring standard	Actual measured size	Score
1	ϕ5.5 mm(2 places)	5 × 2	No score if it is too bad		
2	ϕ5 mm(2 places)	5 × 2	No score if it is too bad		
3	ϕ10 mm(2 places)	5 × 2	No score if it is too bad		
4	(45±0.1) mm(1 place)	5	No score if it is too bad		
5	(30±0.06) mm(1 place)	5	No score if it is too bad		
6	$15^{+0.06}_{0}$ mm	5	No score if it is too bad		
7	$15^{+0.1}_{0}$ mm	5	No score if it is too bad		
8	Safe and civilized production		Points will be deducted for offenders		

(II) Operation steps

Formulate the process, and operate according to the process. The specific operation steps are as follows:

(1) Check the workblank size.

(2) Mark the hole position according to the requirements of the drawing

(3) Use ϕ2 drill bit to drill ϕ2 hole, ϕ4.8 drill bit to drill the bottom hole of ϕ5 hole, and ϕ 5.5 drill bit to drill ϕ5.5 hole.

(4) Ream the hole with ϕ5 reamer.

(5) Countersink the hole with ϕ10 countersink drill.

(III) Relevant precautions

1. When using the drill chuck to clamp the drill bit, use the drill chuck key. Do not knock with flat iron or hand hammer to avoid damaging the chuck and affecting the accuracy of the drill spindle. When clamping the workpiece, clean the clamping surface timely.

2. When drilling, the hand feeding pressure should be controlled by visual inspection and feeling according to the working condition of the drill bit. They should be mastered during practice.

3. The drill bit must be sharpened in time after being blunt.

4. Pay attention to operation safety.

5. Mind the problems that often appear in drilling and the causes, so as to pay attention to them in practice.

Task Five Thread machining

※ Knowledge objectives and skill objectives

(1) Understand the classification, basic dimensions and uses of various threads.

(2) Understand the processing tools for tapping and threading.

(3) Master the processing method and process of tapping.

(4) Master the processing method and process of threading.

I. Basic knowledge cognition of thread

(I) Classification of thread

Thread can be classified as triangle, trapezoid, rectangle, sawtooth and arc thread according to the thread form. it also be classified as single-line thread and multi-line thread according to the number of spiral lines. In addition thread can be classified as cylindrical thread and taper thread according to the shape of the thread body.

Triangular thread commonly used and processed by fitters. There are two kinds of triangular thread: metric type and imperial type. The thread angle of metric triangular thread is 60°, which can be divided into two types: coarse thread and fine thread.

(II) Basic dimensions of thread

1. Elements of the thread

Elements of the thread include thread form, diameter, number of line, pitch and direction of rotation.

(1) The thread form refers to the profile shape of the thread on the face that passes through the thread axis, including triangle, rectangle, trapezoid, arc and sawtooth.

(2) Diameter.

◇ The major diameter (*D, d*) is the diameter of the imaginary cylindrical surface (or taper surface) which coincides with the top of the external thread or the bottom of the internal thread, also named the nominal diameter.

◇ The minor diameter (D_1, d_1) is the diameter of the imaginary cylindrical surface (or taper surface) which coincides with the bottom of the external thread or the top of the internal thread.

◇ The intermediate diameter (D_2, d_2) is between the major diameter and the minor diameter. Assume that there is a cylindrical surface (or taper surface) and in its axial section, the tooth width and groove width on the plain line are equal, and the diameter of the imaginary cylindrical surface is intermediate diameter.

(3) The number of lines (*n*) refers to the number of spiral lines on the thread. Threads can be divided into single-line, double-line and multi-line types. The threads formed along one axis is named single-line thread; the threads formed along two or more spiral lines and equidistant in the axial direction is named multi-line thread.

(4) The pitch (*P*) is the axial distance between two corresponding points of two adjacent teeth on the intermediate diameter line. Lead (P_h) refers to the axial distance between two corresponding points of two adjacent teeth on the intermediate diameter line. For single-line thread, $P = P_h$; for multi-line thread, $P_h = nP$.

(5) The direction of rotation can be divided into two types: left-hand and right-hand. The thread screwed in clockwise is named right-hand thread; the thread screwed in counterclockwise is named left-hand thread. When judging the direction of rotation, the thread rising from left to right is right-handed thread; the thread rising from right to left is left-handed thread.

2. Thread code

The thread code is mainly used to reflect the basic elements of the thread. The expression order of the standard thread code is: the thread form, nominal diameter × pitch (lead or number of lines), accuracy grade and direction of rotation.

(1) The code of ordinary thread is as follows: ordinary thread with coarse teeth is represented by letter M and nominal diameter; ordinary thread with fine teeth is represented by letter M and nominal diameter multiplied by pitch. When the thread is left-handed, add “LH” after the thread code.

(2) Trapezoidal thread code expressed by “Tr”. Single-line trapezoidal thread is represented by Tr and nominal diameter multiplied by pitch; multi-line trapezoidal thread is represented by “Tr” and nominal diameter multiplied by lead; when the thread is left-handed, “LH” should be added.

(3) The code of sawtooth thread is indicated by “B”. Single-line sawtooth thread is represented by “B” and nominal diameter multiplied by pitch; multi-line sawtooth thread is represented by “B” and nominal diameter multiplied by lead; when the thread is left-handed, add “LH” after the thread code.

Examples of ordinary thread, trapezoidal thread and sawtooth thread codes are shown in Table 2-13.

Table 2-13 Examples of thread code

Thread code	**Code description**
M24	Coarse common thread with nominal diameter of 24 mm
M24 × 1.5	Fine common thread with a nominal diameter of 24 mm and a pitch of 1.5 mm
M24 × 1.5LH	Left-handed fine common thread with a nominal diameter of 24 mm and a pitch of 1.5 mm
Tr40 × 7	Single-line trapezoidal thread with a nominal diameter of 40 mm and a pitch of 7 mm
Tr40 × 14(P7)LH	Left-handed multi-line trapezoidal thread with nominal diameter of 40 mm, lead of 14 mm, and pitch of 7 mm
B40 × 7	Single-line sawtooth thread with a nominal diameter of 40 mm and a pitch of 7 mm
B40 × 14(P7)LH	Left-handed multi-line sawtooth thread with nominal diameter of 40 mm, lead of 14 mm, and pitch of 7 mm

II. Tapping

In addition to mechanical processing, which are commonly used triangular thread workpiece can also be obtained by tapping and threading. Tapping is to use a tap to process the internal thread on the cylindrical surface of the workpiece.

(I) Taper and tap wrench

1. Taper

Taper is a forming tool used to process the internal thread with smaller diameter. It is generally made of alloy tool steel 9SiGr and hardened by heat treatment, as shown in Fig. 2-47. Generally, a set of tapers M6–M24 includes the head taper and the second taper; a set of tapers below M6 or above M24 include the head taper, the second taper and the third taper.

Each taper consists of machine part and handle. The machine part consists of cutting part and calibration part. There are several (generally three or four) chip holding grooves in the axial direction, which correspondingly form several blades (cutting edges) and rake angles. The cutting part (that is the incomplete tooth part) is an important part of cutting thread and it is often ground into a cone to distribute the cutting load on several cutter teeth. The angle of the head taper is relatively small, with 5-7 teeth; the angle of the second cone is larger, with 3-4 teeth. The calibration part has complete teeth, which are used to smooth the thread and guide the tap to move in the axial direction. The shank has a square head, which is used to coordinate with the hinge to transmit torque.

Fig. 2-47 Tapers

2. Tap wrench

The tap wrench is a tool used to hold the taper, as shown in Fig. 2-48. The tap wrench is divided into two types: fixed type and adjustable type.

The square hole size and the plate length of the fixed tap wrench should meet certain specifications, so that the tap will not be excessively stressed to prevent it from being broken. It is generally used when tapping below M5.

The size of the square hole of the adjustable tap wrench can be adjusted by rotating the shank, so as to clamp tapers of different sizes. The length of the tap wrench should be selected according to the size of the taper, so as to control the torque during tapping and prevent the taper from being broken due to improper force application.

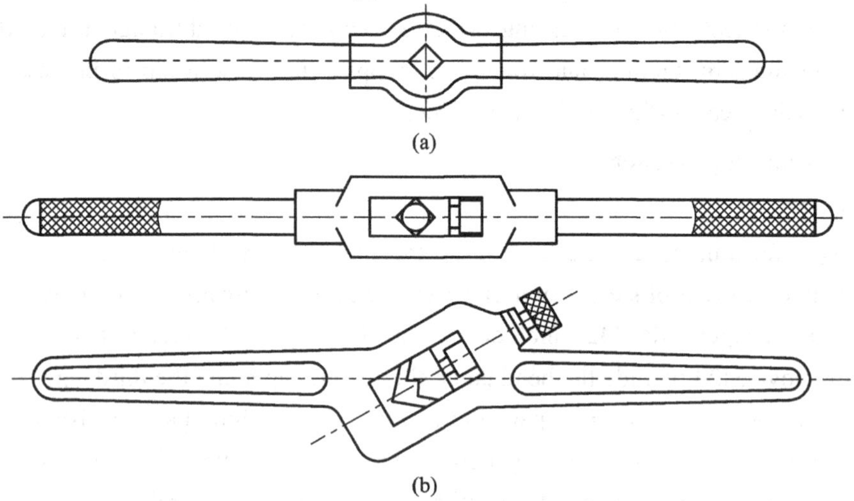

Fig. 2-48 Tap wrench

(II) Process of tapping

1. Determination of the diameter of the bottom hole

In the process of tapping, the cutting edge not only cuts the metal, but also squeezes the metal, which causes the metal to bulge and flow to the tooth tip. Therefore, before tapping, the hole diameter (bottom hole) drilled should be larger than the inner diameter of the thread. The diameter of the bottom hole can be found in the manual or calculated according to the following empirical formula.

Brittle materials (cast iron, bronze, etc.): $d_0 = D - 1.1P$

Shaping materials (steel, copper, etc.): $d_0 = D - P$

In the above, d_0 represents drilling hole diameter.

2. Determination of drilling depth

When tapping the thread of a blind hole (non-through hole), the depth of the hole must be greater than the length of the thread because the tap cannot tap to the end. The depth of the blind hole can be calculated according to the following formula: the depth of the hole is equal to the depth of the required thread plus $0.7D$.

3. Determination of hole chamfering

Before tapping, chamfering should be carried out at the hole of the drilling hole to facilitate the positioning and cutting in of the tap. The depth of the chamfer is greater than the pitch of thread.

(III) Key points and precautions of tapping

(1) According to the specification of the thread hole on the workpiece, the taper should be selected correctly. The head taper first and then the second taper, and the order can not be violated.

(2) When the workpiece is clamped, the center of the hole should be perpendicular to the jaw to prevent the tapping from being crooked.

(3) When tapping the thread with a head taper, after screwing in 1 to 2 turns, check whether the taper is perpendicular to the end face of the hole (It can be inspected visually or with a square in two mutually perpendicular directions). When the cutting part has been cut into the workpiece, it should be reversed 1/4 turn every 1 to 2 turns so that the chips will break off. At the same time, no pressure can be applied (that is, only turning without pressure), so as to prevent the taper from being broken or the tapped thread teeth being too thin.

(4) The tapping of the internal thread of steel parts should be lubricated with engine oil, which can make the thread smooth, save efforts and extend the service life of the tap. It is possible for the tapping of the internal thread of the cast iron to be carried out without lubricant or kerosene. The tapping of the internal thread of the aluminum or aluminum alloy, or copper, can be lubricated with emulsion.

(5) Do not blow the chips directly with your mouth to prevent the chips from flying into the eyes.

III. Threading

(I) Threading die and die handle

1. Threading die

Threading die is a tool for machining external thread, which is made of alloy tool steel 9SiGr and hardened by heat treatment. Its shape is like a round nut, but there are 3 to 4 chip removing holes drilled on it to form a blade, as shown in Fig. 2-49.

Fig. 2-49 Threading die

The threading die is composed of a chip part, a positioning part and a chip removing hole. There is a 40° taper at both ends of the thread hole of the round die, which is the cutting part of the die. The positioning part plays the role of polishing. The outer circle of the threading die has a deep groove and four cone pits, which are used for positioning and fastening the die.

2. Die handle

The die handle is a tool for clamping the threading die and transmitting torque, as shown in

Fig. 2-50.

Different die handle should be selected for threading dies with different outer diameters.

Fig. 2-50 Die handle

(II) The process of threading

1. Determination of the diameter of the round rod

The same as tapping thread, threading has the function of cutting and extruding metal. Therefore, the diameter of the round rod must be checked before threading. The diameter of round rod should be slightly smaller than the nominal diameter of thread, which can be calculated by referring to table or empirical formula.

Empirical formula: diameter of round rod is equal to outer diameter of thread minus(0.13–0.2) times pitch

2. Determination of Chamfering at the end of the round rod

The end of the round rod should be chamfered before threading, so that the threading die can be easily aligned with the center of the workpiece, and easily cut in at the same time. The chamfer length should be greater than one pitch, and the bevel angle should be 15°～30°.

(III) Key points and precautions for threading

The key points and precautions for threading are as follows:

(1) The chip in the threading die chip removal groove and the thread should be cleaned before threading every time.

(2) Check the diameter of the round rod and the end chamfer before threading.

(3) The cutting torque is very large when threading, and it is easy to damage the processed surface of the round rod. Therefore, a hardwood V-groove liner or a thick copper plate should be used as a protective sheet to clamp the workpiece. The length of the workpiece protruding from the jaws should be as short as possible without affecting the required length of the thread.

(4) When threading, the end face of the threading die should be perpendicular to the round rod, and the force should be uniform during operation. When starting to rotate the die, apply a little pressure. After threading 3 to 4 teeth, one can only rotate without pressing, and often reverse

to break the chips.

(5) Lubricate with engine oil when threading the steel rod.

IV. Practice of tapping and threading

The drawing of tapping and threading is shown in Fig. 2-51.

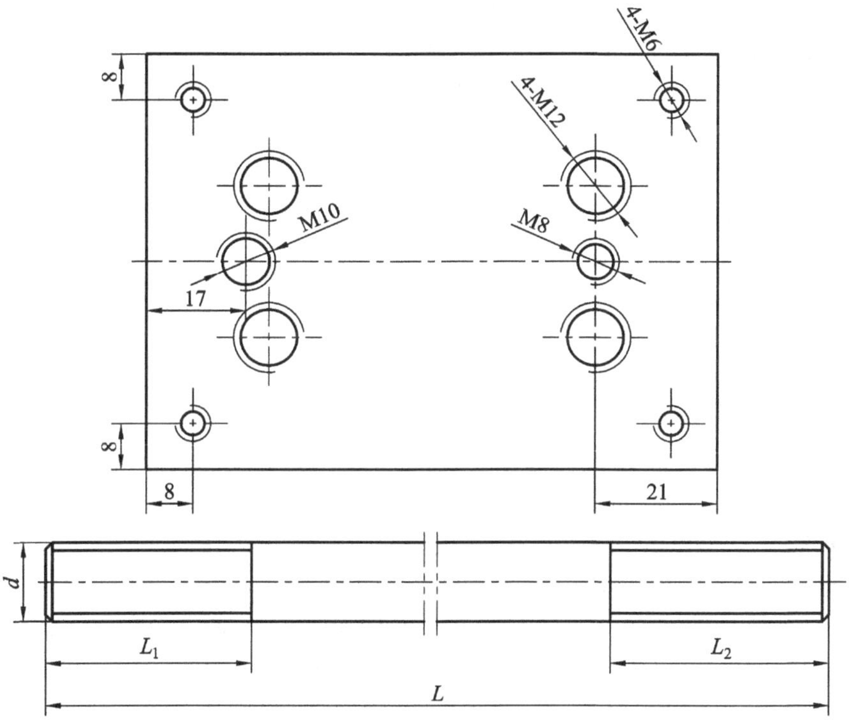

No.	*d*/mm	*L*/mm	L_1/mm	L_2/mm
1	8	100	20	30
2	10	150	20	40
3	12	200	20	50

Fig. 2-51　Thread processing

(I) Preparations before operation

Analyze the drawings and clarify the requirements: determine the machining and measuring tools as needed, and determine the material according to the drawing; Formulate processing technology and make preparations before processing.

Select of machining tools and measuring tools: according to the drawing requirements, choose machining tools and measuring tools reasonably. See Table 2-14 for details.

Table 2-14 List of machining tools and measuring tools

No.	Name	Specification	Quantity	Remarks
1	Vernier caliper	0-150 mm	1	Precision of 0.02 mm
2	Knife-edge ruler	125 mm	1	Grade 1
3	Height vernier caliper	0-300 mm	As required	Precision of 0.02 mm
4	Square		1	
5	Straight shank twist drill	ϕ5 mm、ϕ6.8 mm、ϕ8.5 mm、ϕ10.3 mm	1 for each	
6	Taper	M6、M8、M10	1 for each	
7	Die	M8、M10、M12	1 for each	
8	Saw bow, Saw blade, Hand hammer, Scribing compass, Marking pin, Sample punch, Steel ruler, Soft jaws, File brush, Brush, Motor oil, Pencil, etc.		As required	

(II) Operation steps

Formulate the process, and operate according to the process. The specific operation steps are as follows:

(1) Check the workblank size.

(2) Mark the hole position line according to the drawing requirements.

(3) Drill corresponding bottom holes with twist drills of ϕ5, ϕ6.8, ϕ8.5 and ϕ10.3.

(4) Tap the thread with tapers of M6, M8, M10 and M12.

(5) Use M8, M10 and M12 die to finish the external threading.

(III) Relevant precautions

The relevant precautions for tapping and threading are as follows:

(1) According to the specification of the thread hole on the workpiece, the tapers should be selected correctly. The head taper first and then the second taper, and the order can not be violated;

(2) When tapping the thread with a head taper, after screwing in 1 to 2 turns, check whether the taper is perpendicular to the end face of the hole (It can be inspected visually or with a square in two mutually perpendicular directions). When the cutting part has been cut into the workpiece, it should be reversed 1/4 turn every 1 to 2 turns so that the chips will break off. At the same time, no pressure can be applied (that is, only turning without pressure), so as to prevent the taper from being broken or the tapped thread teeth being too thinner;

(3) The chip in the die chip removal groove and the thread should be cleaned before threading every time;

(4) When threading, the end face of the die should be perpendicular to the round rod, and the force should be uniform during operation. When starting to rotate the die, apply a little

pressure. After threading 3 to 4 teeth, you can only rotate without pressure, and often reverse to break the chips;

(5) Pay attention to safe operation.

Questions for review

1. What is plane marking? What is three-dimensional marking?
2. What is the marking datum? What are the principles of selecting marking datum?
3. How many methods are there for saw entry? Which one is better?
4. Briefly describe the selection principle of saw blades?
5. Briefly describe the methods of sawing thin sheets?
6. Briefly describe the type and scope of application of files?
7. How many methods are there for filing a plane?
8. Briefly describe the detection method of flatness during plane filing?
9. Briefly describe the structure of the twist drill?
10. What are the main types of drilling machines?
11. Briefly describe the method of marking and drilling?
12. What are the types and characteristics of countersink drills?
13. What are the structure and types of reamer?
14. What are the elements of thread?
15. How is the diameter of the bottom hole calculated when tapping?

Module Three　Simple mechanical assembly

Task One　Assembly of fixed connection

※ Knowledge objectives and skill objectives

(1) Be familiar with various fixed connection assembly tools.

(2) Master the types, characteristics of thread connection and anti-loosening device.

(3) Master the specification code, matching type and selection knowledge of flat keys.

(4) Master the types of pins and application knowledge of pin connections.

I. Common assembly tools for fixed connection

At present, in most factories, assembly work is mostly done by manual labor, and many hand tools are used. In addition to screwdrivers, pliers, files, hand hammers and some measuring and marking tools commonly used by fitters, various wrenches, retaining ring pliers, pin pullers and press machines are often used.

(I) Wrench

Wrench is used to tighten hexagonal and square screws as well as various nuts. It is made of tool steel, alloy steel or malleable cast iron. Its opening is required to be smooth, hard and wear-resistant. Wrenches can be divided into three types: adjustable wrench, special-purpose wrench and special wrench.

1. Adjustable wrench (universal wrench)

The adjustable wrench is composed of wrench body, fixed jaw, movable jaw and worm, as shown in Fig. 3-1. The opening size of adjustable wrench can be adjusted within a certain range. Its specification is expressed by the length and the opening size of the

Fig. 3-1　Adjustable wrench

wrench, but generally the length of the wrench is used as its specification, such as 100 mm, 150 mm, 200 mm, 250 mm, 300 mm, 375 mm, 450 mm, 600 mm, etc.

Pay attention to the following points when using adjustable wrench:

(1) The fixed jaws should be subjected to the main force, otherwise the wrench is easily damaged.

(2) the opening size of the jaws should be suitable for the size of the nut's opposite side spacing, otherwise the nut will be damaged.

(3) For different specifications of nuts (or screws), wrench of the corresponding specifications should be used.

(4) The wrench handle can not be extended arbitrarily, so as not to damage the wrench or screw due to excessive tightening torque.

(5) The working efficiency of the adjustable wrench is not high, and the movable jaw is easy to skew, which often damages the head surface of the nut or screw.

2. Special-purpose wrench

Special-purpose wrench can only pull nuts or screws of one size, which can be divided into the following types according to their different uses.

1) Open-end wrench (solid wrench)

Open-end wrench are used for loading and unloading hexagonal or square-headed nuts or screws, which can be divided into single-headed wrench and double-headed wrench, as shown in Fig. 3-2. Their opening size is compatible with the size of the distance between the opposite sides of the screw and nut, and a set is made according to the standard size.

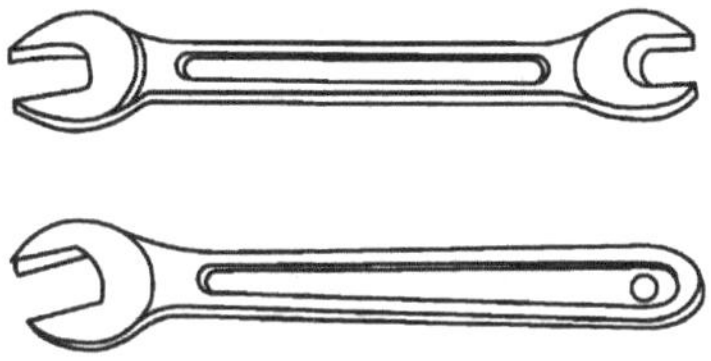

Fig. 3-2 Open-end wrench

2) Integral wrench

There are several types of integral wrench, such as square, hexagonal, and dodecagonal (box wrench, shown in Fig. 3-3). Box wrench are widely used. It can change direction and pull again as long as it turns 30°, so it can work in places with a narrow range.

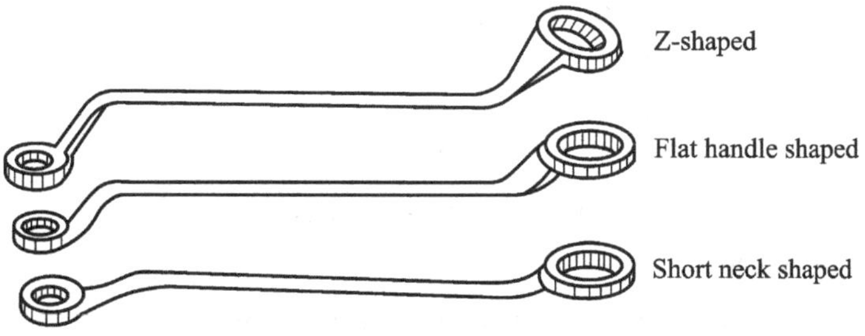

Fig. 3-3 Box wrench

3) Socket wrench set

Socket wrench set is composed of a set of box sockets and wrench handles of varying sizes. The square tenon of the wrench handle can be inserted into the square hole of the box socket to work. Among them, the bow-shaped handle can be continuously rotated, and the ratchet handle can be continuously moved back and forth, so it is convenient to use and has high work efficiency.

4) Internal hexagonal wrench

The inner hexagonal wrench is used to tighten the inner hexagonal screw. This kind of wrench is a complete set, as shown in Fig. 3-4. The inner hexagonal wrench can tighten the inner hexagonal screw of M3–M24, and its specification is expressed by the size of the hexagonal opposite side spacing.

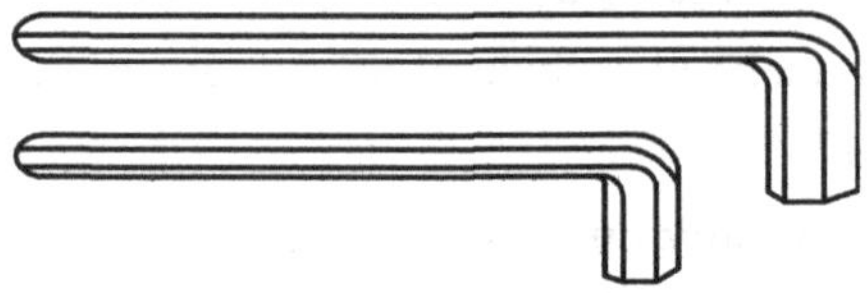

Fig. 3-4 Inner hexagonal wrench

3. Special wrench

Special wrench is made according to some special requirements.

1) Ratchet wrench

As shown in Fig. 3-5, ratchet wrench is suitable for narrow places. During work, when the handle is turned forward, the pawl1# enters the gap of the inner hexagonal socket (ratchet wheel) under the action of the spring, and the sleeve rotates accordingly. When the handle is turned in the reverse direction, the pawl will withdraw from the gap of the sleeve to slip under the action of the inclined surface, so the nut will not rotate reversely accordingly.

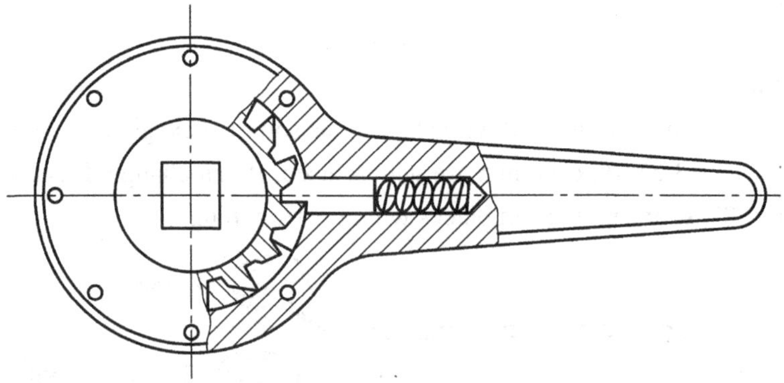

Fig. 3-5 Ratchet Wrench

2) Pneumatic wrench

Pneumatic wrenches are powered by compressed air and are suitable for tightening and disassembling the thread connection in the installation of mass production of automobiles and tractors, as shown in Fig. 3-6. Pneumatic wrenches can select suitable torque bars according to the size of the bolts and the required torque value to achieve different fixed torque requirements. Pneumatic wrenches are especially suitable for continuous production of mechanical assembly lines, which can improve assembly quality and efficiency, and reduce labor intensity.

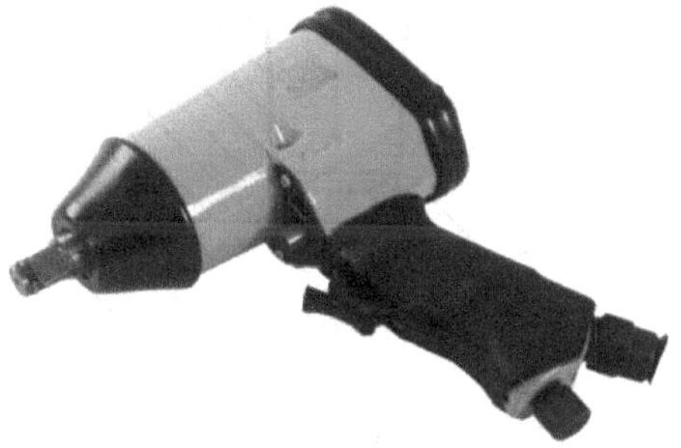

Fig. 3-6 Pneumatic wrench

(II) Retaining ring plier

The retaining ring plier is specially used for assembling and disassembling elastic retaining ring. As the retaining ring type is divided into two types:hole type and shaft type, and the installation position is different, the retaining ring plier can be divided into straight-nozzle retaining ring plier and curved-nozzle retaining ring plier, as shown in Fig. 3-7.

Fig. 3-7 Retaining ring plier

(III) Pin puller

The pin puller is specially used to remove the tapered pin with thread ends. For the taper pin with internal thread as shown in Fig. 3-8 or the taper pin that can only be assembled and disassembled on one side, it is difficult to disassemble. The pin puller shown in Fig. 3-9 is often used for disassembly. When disassembling, first screw the thread of the pin puller into the internal thread of the pin, and then quickly slide the slider on the pin puller outward to generate an outward impact force to pull out the tapered pin.

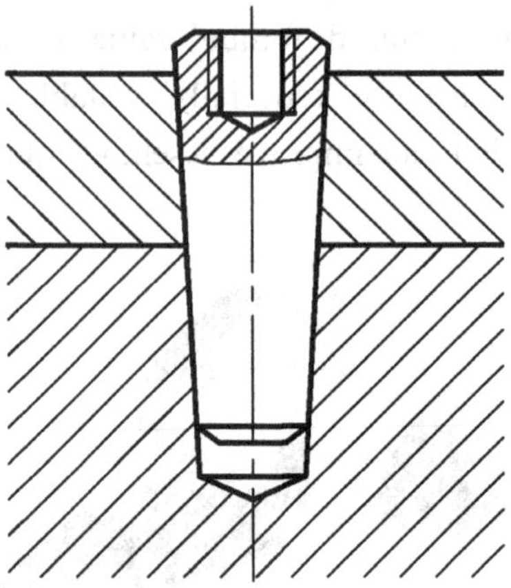

Fig. 3-8 Tapered pin with internal thread

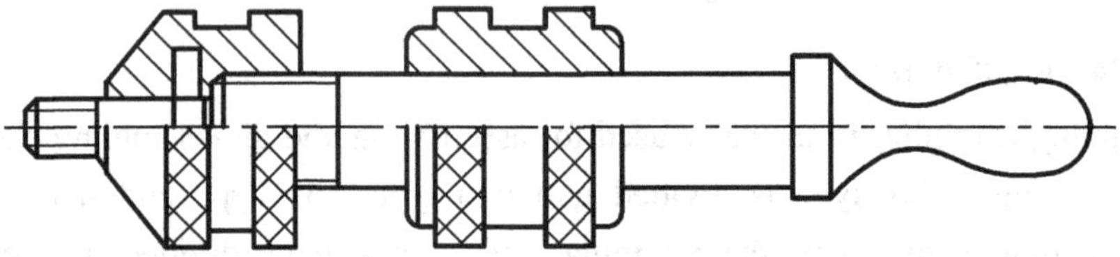

Fig. 3-9 Pin puller

II. Threaded connection

(I) Classification of thread

1. Triangular thread

Triangular thread has large thread angle, good self-locking performance, thick root and high strength, so it is often used for connection. The commonly used triangular thread include ordinary thread, inch thread and cylindrical pipe thread.

1) Ordinary thread

In the national standard, the triangular metric thread with thread angle $\alpha = 60°$ is called ordinary thread. The major diameter d is the nominal diameter, and the thread feature code is *M*. Ordinary thread is divided into two types: coarse thread and fine thread, but they share the same code. Generally, coarse thread is used for connection. When the major diameter of the thread is the same, the pitch and thread height of the fine thread are smaller than those of the coarse thread, so the fine thread is suitable for connecting thin-walled parts.

Ordinary threads are generally used to fasten bolts and nuts of various equipment, which can be divided into metric system and imperial system.

2) Inch screw thread

For the Inch screw thread, the thread angle $\alpha = 55°$. It is recorded in imperials, and the pitch is expressed by the number of teeth per inch. There are also coarse teeth and fine teeth. Imperial screw thread is mainly used in the United Kingdom, the United States and other countries, and it is generally only used in repairs in China.

3) Cylindrical pipe thread

The thread angle of the cylindrical pipe thread is 55° and the top of the thread is in arc shape. There is no radial gap between the screw threads, and the tightness is good. The nominal diameter is the nominal diameter of the pipe, which is widely used in the connection of water, gas, lubrication and other pipeline systems.

2. Rectangular thread

The thread form of rectangular thread is square, the thread angle $\alpha = 0°$. The tooth thickness is half of the pitch. The equivalent coefficient of friction is small and the efficiency is high, but the root strength is low, and the axial gap caused by thread wear is difficult to compensate. Its alignment accuracy is low, and the finishing is difficult. Therefore, this kind of thread has been seldom used.

3. Trapezoidal thread

The thread form of trapezoidal thread is isosceles trapezoid with a thread angle $\alpha = 30°$ and lower efficiency than that of rectangular thread. However, it is easy to process, has good alignment accuracy and high root strength. When the split nut is used, the gap caused by wear can be eliminated, so it is widely used in screw drive.

4. Sawtooth thread

The flank angle of the sawtooth thread working face is 3° and the flank angle of non-working face is 30°. It has the advantages of high efficiency of rectangular thread and high root strength of trapezoidal thread, but it can only bear unidirectional load and is suitable for screw transmission with unidirectional load. Serrated threads have high root strength and are used for transmission screws with unidirectional force , such as screw presses, Jacks, etc.

(II) Basic types of threaded connection

Threaded connection is a kind of detachable fixed connection, which has the advantages of simple structure, reliable connection, convenient assembly and disassembly, so it is widely used in machinery (as shown in Fig. 3-10). The screw connection is suitable for the connection where the connected parts are rarely removed.

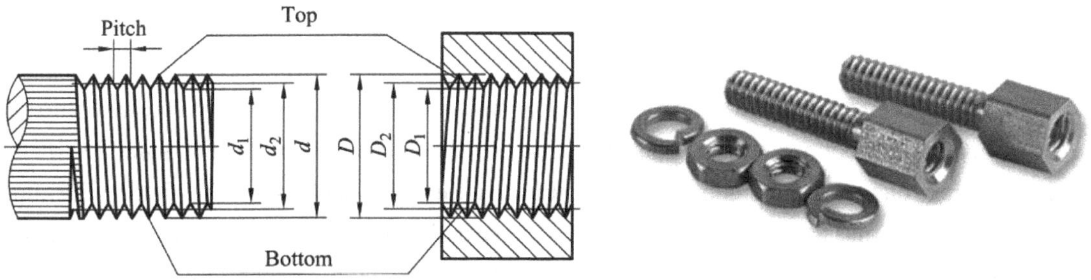

Fig. 3-10 Threaded connection

1. Bolted connection

Bolted connection is used for the connection between workpieces which require convenient assembly and disassembly, and the holes of the connected parts are not cut with threads. As shown in Fig. 3-11 (a), it is an ordinary bolted connection with a gap between the bolt and the hole. It is the most widely used due to its simple processing and low cost. As shown in Fig. 3-11 (b), the reamed hole is connected with bolt, and the hole on the connected part is processed with a high-precision reamer, and the transition fit is generally adopted between the bolt rod and the hole, which is mainly in places where the bolt is required to bear the lateral load or the bolt rod is required to accurately fix the relative position of the connected parts.

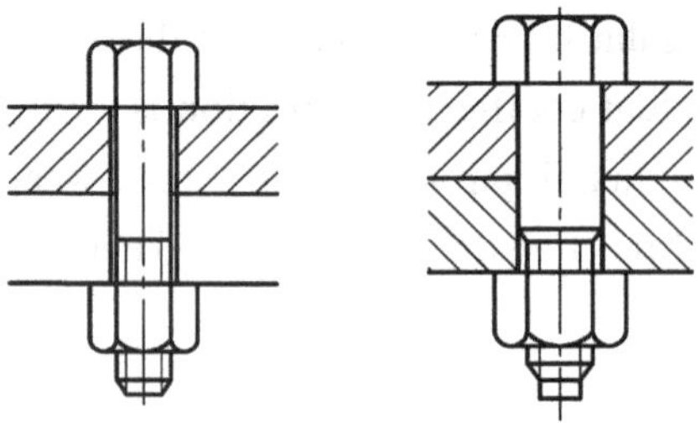

(a) (b)

Fig. 3-11 Bolted connection

2. Stud connection

Stud connection refers to the connection of stud with threads at both ends, one end of which is screwed into and fixed in the threaded hole of the thicker connected part, and the other end passes through the through hole of the thinner connected part (see Fig. 3-12). The stud connection is suitable for the occasions when the connected parts are thick, the structure is compact and the assembly and disassembly are frequent.

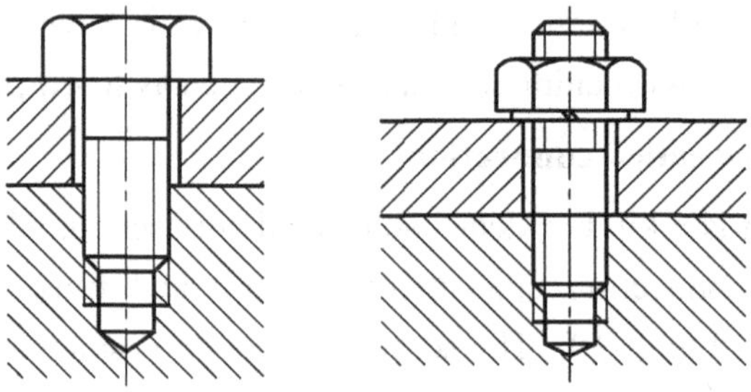

(a) (b)

Fig. 3-12 Stud connection

3. Screw connection

The screw connection refers to the connection when the screw is directly screwed into the threaded hole of the connected part, and the structure is relatively simple. It is suitable for the occasions when one of the connected parts is thick or the other end cannot be fitted with nuts. However, frequent disassembly and assembly will wear the threaded hole and cause premature failure of the connected parts, so it is not suitable for frequent disassembly and assembly occasions.

4. Set screw connection

Set screw connection is to screw the set screw into the threaded hole of one part, and its end is pressed against the surface of another part (see Fig. 3-13), or pushed into the corresponding pit. Set screw connection is often used to fix the relative position of two parts, and can transfer little force or torque.

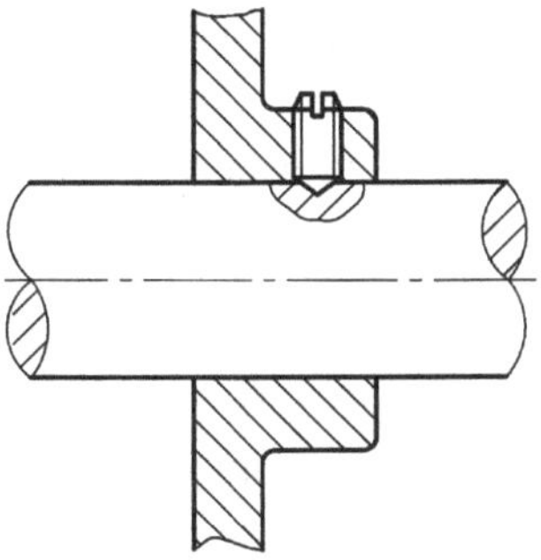

Fig. 3-13 Screw connection

(Ⅲ) Pre-tightening of threaded connection

In order to make the connection firm and reliable, there must be enough torque when tightening the thread. The torque of threaded connection is generally controlled by the assembler based on the experience, and the tightening torque should be appropriate. For threaded connection that require pre-tightening force, the pr-tightening force requirement can be found in the process documents.

(Ⅳ) Anti-loosening device of the threaded connection

The fitting accuracy must be ensured when the thread is connected. In order to prevent the loosening of the thread under the impact load, a reliable anti-loosening device must be provided during threaded connection(see Fig. 3-14).

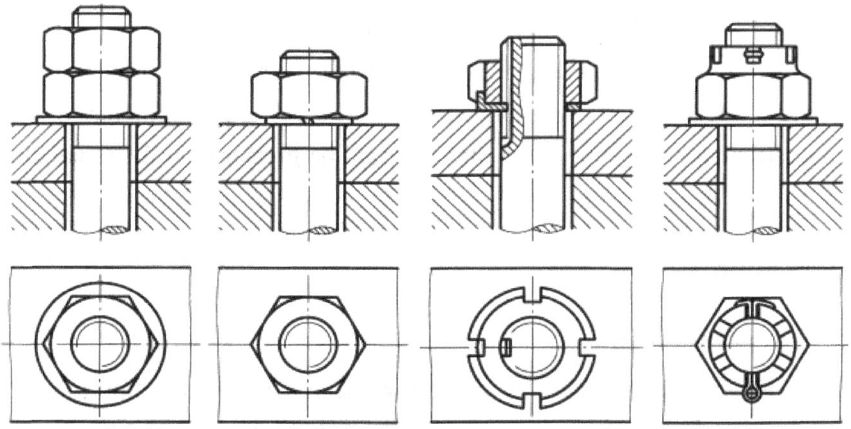

(a) Counter nut (b) Spring washer (c) Tap washer (d) Split pin and slotted nut

Fig. 3-14 Anti-loosening device of the threaded connection

(1) Counter nut. There are main nut and the auxiliary nut here. First tighten the main nut to

the predetermined position, and then tighten the auxiliary nut. Because two nuts are used here, the structure size and quality are increased, and it is generally used in low-speed heavy load or relatively stable occasions.

(2) Spring washer. This kind of anti-loosening device is easy to scratch the surface of the nut and the connected part. At the same time, the nut is easy to deflect due to the uneven distribution of the elastic force. With the simple structure, it is generally used in situations that is stable an disassembled infrequently.

(3) Tap washer. The ear parts of the washers are respectively in close contact with the connecting piece and the hexagonal screw or nut to prevent loosening. It is mainly used in situations where the connecting part can hold the bending ears.

(4) Split pin and slotted nut. Use split pin to lock the nut directly on the bolt. It is anti-loosing and reliable, but the position of the pin hole in the screw is not easy to match with the slot of the best locking position of the nut. It is mainly used in the occasions of variable load and vibration.

(V) Assembly points of bolted or screw connection

(1) Before installation, check whether the joint surface between the workpiece and nut, bolt (nail) is flat, whether the axis of screw hole is perpendicular to the joint surface, whether the thread is damaged, and whether the tooth shape is correct. If it does not meet the requirements, it cannot be connected before processing.

(2) For threaded connection with torque requirements, torque wrench or force limiting devices shall be used. Never increase the length of the wrench to avoid exceeding the allowable load when tightening the thread.

(3) When the same workpiece is connected by multiple threads, it should be tightened gradually and evenly several times in a symmetrical order to avoid deformation of the workpiece.

(4) After the nut is tightened, the length of the exposed part of the bolt shall not be less than one screw pitch, and no more than the thickness of 1 nut.

(5) When the connecting screw is connected to the threaded plate under the radial shear external force, the connection must be stable and tight. The thread pre-tightening force multiplied by the friction coefficient between the two plates should be greater than the radial shear external force.

III. Key connection

(I) Type of key connection

Key is a standard part used to connect the transmission parts and transmit torque. According to the different structural characteristics and uses, the key connection can be divided into four types: flat key connection, woodruff key connection, taper key connection and tangential key connection. Among them, the fixed connection includes woodruff key connection, taper key connection, tangential key connection, and ordinary flat key connection and thin flat key connection in flat key connection (see Fig. 3-15). The guide key and sliding key in flat key connection belong to movable connection, and the flat key connection can not realize the axial positioning of parts on the shaft.

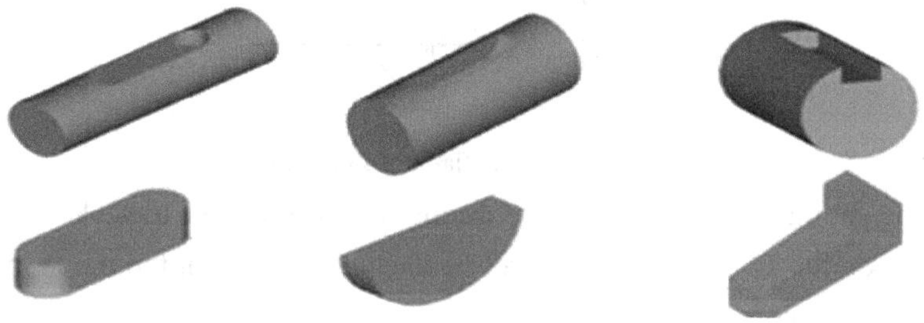

(a) Flat key connection (b) Woodruff key connection (c) Taper key connection

Fig. 3-15 Key connection

(II) Assembly of key connection

The key connection is used to connect the shaft and the parts on the shaft to realize the circumferential fixation of the parts and transmit torque. The key connection has the advantages of simple structure, reliable work, convenient assembly and disassembly, and is widely used in mechanical connection. According to different structural characteristics and uses, the key connection can be divided into loose key connection and tight key connection.

1. Assembly of loose key connection

There are three types of keys used for loose key connection: ordinary flat key, guide key and woodruff key. The key is placed in the shaft groove and tightly attached to the groove bottom. There is a gap of 0.1 mm between the length direction and the shaft groove, and a gap of 0.3–0.5 mm between the key top and the hub groove.

(1) Common flat key connection: P9 / h9 or N9 / h9 is used to fit the key and shaft groove. The fit of the key and hub groove is Js9 / h9 or P9 / h9, that is, the key is fixed on the shaft and hub, as shown in Fig. 3-16.

(2) Woodruff key connection: It is generally used for light load, suitable for the tapered end of the shaft, as shown in Fig. 3-17.

(3) Guide flat key connection: H9 / h9 is used to fit the key with the shaft groove and fixed on the shaft with screws, D10 / h9 is used to fit the key with the hub, and the parts on the shaft can move axially, as shown in Fig. 3-18.

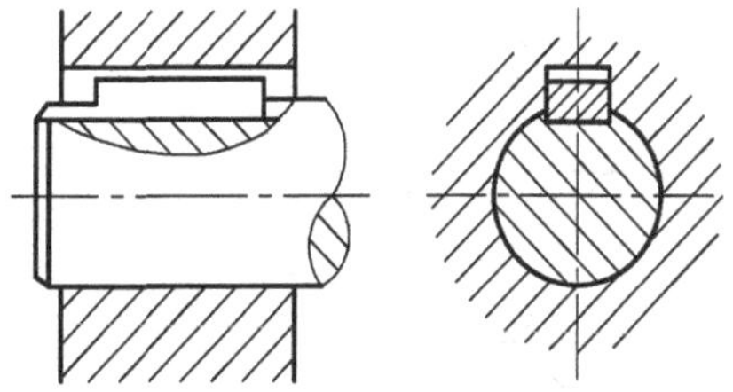

Fig. 3-16 Common flat key connection

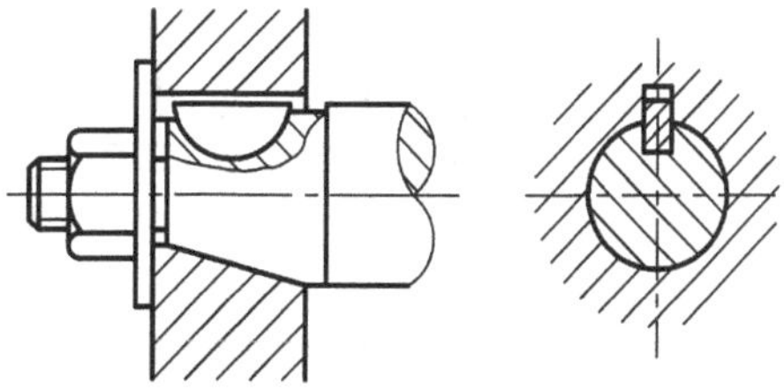

Fig. 3-17 Woodruff key connection

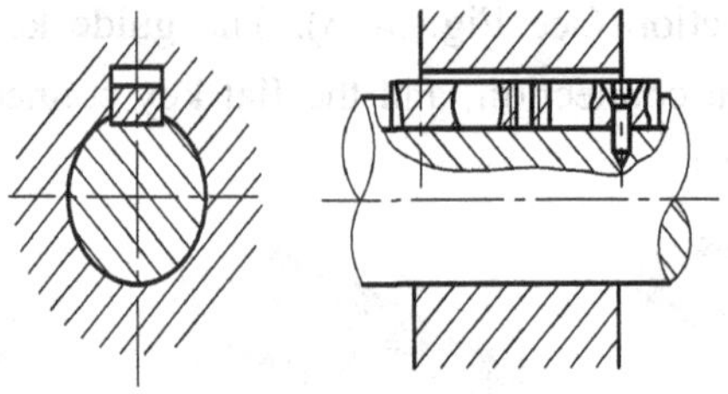

Fig. 3-18 Guide flat key connection

The following should be paid attention to during the assembly of loose key connection:

(1) The burrs on the key and keyway shall be cleaned to prevent too much interference after fitting and damaging the correctness of the fitting.

(2) For important key connection, check the straightness of the key and the symmetry and parallelism of the keyway to the axis before assembly.

(3) Try to match the head of the key with the shaft groove, so that the key can be tightly embedded in the shaft groove (for common flat key and guide flat key).

(4) Add engine oil to the fitting surface, press the key into the shaft groove with copper rod or bench vice (soft jaws should be added) and make good contact with the groove bottom.

(5) During trial assembly and installation of kits (gears, pulleys, etc.), there should be a gap between the non-matching surface of the key and the keyway, so that the shaft and the kit can meet the requirements of coaxiality. The assembled kit can not swing left or right on the shaft, otherwise, it is easy to cause impact and vibration.

2. Assembly of tight key connection

Tight key connection mainly refers to taper key connection, as shown in Fig. 3-19. Tight key connection is divided into common taper key connection and hook head taper key connection. The upper and lower sides of the taper key are working surfaces. The upper surface of the key and the bottom surface of the hub groove have a slope of 1 ∶ 100, and there is a certain gap between the key side and the keyway. When assembling, it needs to be driven in, and the torque can be transmitted by the taper action, and it can also axially fix the parts and bear the unidirectional axial force. Because the parts on the shaft are easy to produce eccentricity and skew when they fit with the shaft, they are mostly used in occasions where the neutral requirements are not high and the rotation speed is low. The hook-head taper key is used when the key cannot be punched out from the other end.

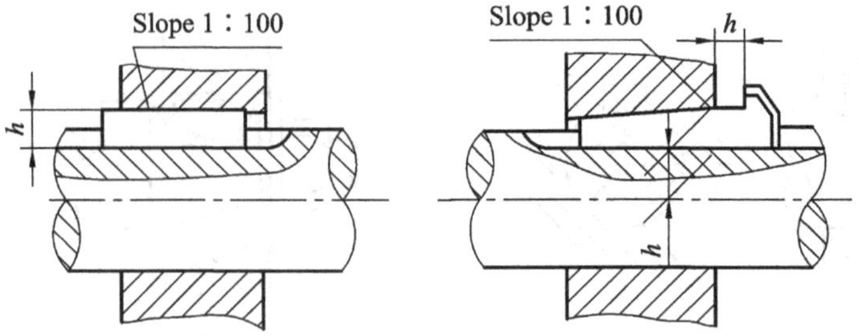

Fig. 3-19 Tight key connection

The following points of the assembly of the tight key connection should be minded:

(1) The inclination of the taper key should be consistent with that of the hub groove, otherwise, the kit will skew and the connection strength will be reduced.

(2) There should be a certain gap between the taper key and both sides of the slot.

(3) For the hook-head taper key, the hook head should not be close to the end face of the kit, and a certain distance must be left for the convenience of disassembly.

(4) When assembling the taper key, check the contact of the upper and lower surfaces of the taper key and the shaft groove or hub groove with the coloring method. If there is poor contact, the key groove can be trimmed with a file or scraper. After passing the test, tap it in gently until it is securely fastened in the circumferential and axial directions.

IV. Pinned connection

(I) Types of pinned connection

Pinned connection is a kind of connection that uses pin or pin shaft to connect two or more parts together. Pinned connection not only plays the role of connection, but also plays the role of positioning in the mechanism. According to the pin structure, pinned connection can be divided into three types: cylindrical pinned connection, tapered pinned connection and split pinned connection (see Fig. 3-20). According to the functional requirements, pinned connection can be divided into three types: fixed pinned connection, positioning pinned connection and hinge pinned connection.

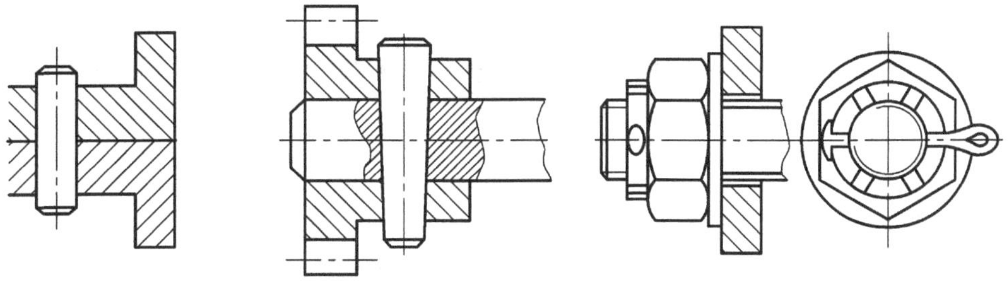

(a) Cylindrical pinned connection (b) Tapered pinned connection (c) Split pinned connection

Fig. 3-20 Pinned connection

(II) Assembly of pinned connection

The structure of pinned connection is simple, the positioning and connection are reliable, and the assembly and disassembly are convenient. It not only plays the role of connection, but also plays the role of positioning and safety insurance, as shown in Fig. 3-21.

1. Assembly of cylindrical pinned connection

The cylindrical pinned connection are manufactured according to four deviations of n6, g6, h8 and h9. In most cases, there is a small amount of interference between the cylindrical pinned connection and the pin hole to ensure the tightness and accuracy of the connection or positioning. Cylindrical pinned connection with interference fit is not suitable for frequent assembly and disassembly. For the cylindrical pinned connection positioning structure that needs to be assembled and disassembled frequently, in order to facilitate assembly and disassembly, one of the two positioning pin holes generally adopts clearance fit.

2. Assembly of taper pinned connection

Taper pinned connection has a taper of 1∶50 and is accurate in positioning. It can be assembled and disassembled for many times without reducing the positioning accuracy and is widely used. Taper pinned connection is expressed in terms of small end diameter and length. During assembly, it is generally measured by trial assembly method. It is appropriate when the pin can be inserted into about 80% of its length.

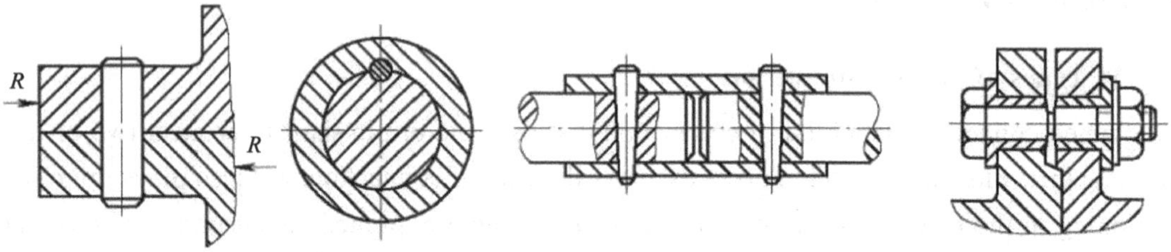

(a) For positioning (b) For connection (c) For Safety insurance

Fig. 3-21 Assembly of pinned connection

Task Two Assembly of belt drive mechanism

※ Knowledge objectives and skill objectives

(1) Understand the composition, principle, classification and characteristics of belt drive.

(2) Master the tension and adjustment of V-belt drive mechanism.

(3) Master the assembly method and adjustment skills of belt drive.

I. The composition and principle of belt drive

Belt drive is a common mechanical drive, which relies on the friction or meshing force between the belt and the pulley to transmit power.

1. The composition of the belt drive

The composition of the belt drive is shown in Fig. 3-22.

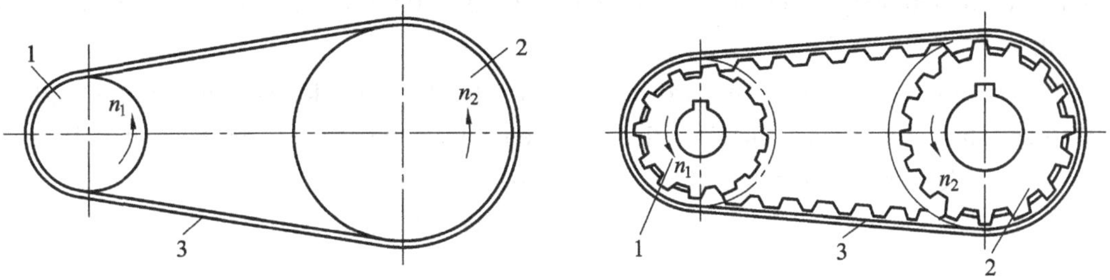

(a) Friction type belt drive (b) Meshing type belt drive

1-Belt pulley(driving wheel); 2-Belt pulley(driven wheel); 3-Flexible belt.

Fig. 3-22 The composition of the belt drive

2. Working principle of belt drive

The belt drive takes the belt tensioned on two or more wheels as the intermediate flexible part, and transmits the motion and /or power by the friction (meshing force) generated between the belt and the belt pulley.

3. Transmission ratio i of belt drive

The transmission ratio of mechanism is the ratio of instantaneous input speed to the output speed in the mechanism.

The transmission ratio of the belt drive is the ratio of driving wheel speed n_1 to the driven wheel speed n_2.

II. Types and characteristics of the belt drive

1. Types of belt drive

According to the section shape of belt, the belt drive can be divided into five types: flat belt, V-belt, round belt, poly V-belt and synchronous belt, as shown in Fig. 3-23. Among them, V-belt drive is the most widely used.

(a) Flat belt (b) V-belt (c) Round belt (d) Poly V-belt (e) Synchronous belt

Fig. 3-23 Types of belt drives

2. Characteristics of belt drive

Compared with gear drive, belt drive has the advantages of stable operation, low noise, simple structure, no lubrication, vibration absorption, easy manufacturing, overload protection, large transmission center distance and so on. The disadvantages of belt drive are inaccurate transmission ratio, low drive efficiency and short belt life.

III. Assembly requirements of V-belt drive mechanism

(1) The belt pulley must be installed correctly. It is generally required that the radial circular

runout is (0.0025−0.005) D, the end face runout is (0.0005−0.001) D, and D is the diameter of the belt pulley.

(2) The middle planes of the two belt pulleys should coincide. The inclination angle and axial offset shall not exceed the specified requirements, generally 1°, or the belt will fall off easily or the side wear of the belt will be accelerated.

(3) The smoothness of the working surface of the belt pulley should be appropriate. Ra is generally not less than 3.2 μm.

(4) The wrap angle attached to the belt pulley should not be too small. For the V-belt, the wrap angle should not be less than 120°, or it will slip easily.

(5) The tension of the belt pulley must be proper. If the tension is too small, a certain amount of power can not be transmitted; if the tension is too large, the belts, shafts and bearings are easy to wear, and the stability of the drive is reduced. Therefore, the proper tension is an important factor to ensure that the belt drive can work normally.

IV. Adjustment of the tension

The belt drive mechanism is designed with a tension device to adjust the tension. The pulling force can be restored to the original requirement by adjusting the center distance between the two shafts.

1. Inspection of the tension

As shown in Fig. 3-24, in V-belt drive, it is stipulated that under the action of measured load W, the tension that causes 1.6 mm deflection at the midpoint by every 100 mm of the tangent point of the belt and two wheels is the appropriate value, that is, under the action of measured load W, the deflection $y=\frac{1.6}{100}t$ mm, and t is the distance between the two tangent points of V-belt.

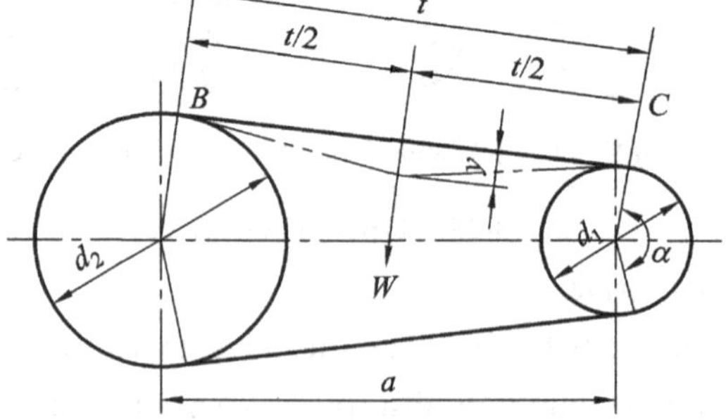

Fig. 3-24 Inspection of the tension

If the measured deflection is greater than the calculated value, the tension is less than the specified value; otherwise, the tension is greater than the specified value.

2. Adjustment of the tension

As shown in Fig. 3-25, there are two ways to adjust the tension:

(1) Adjust the center distance between two belt pulleys.

(2) Adjust with tensioning wheel.

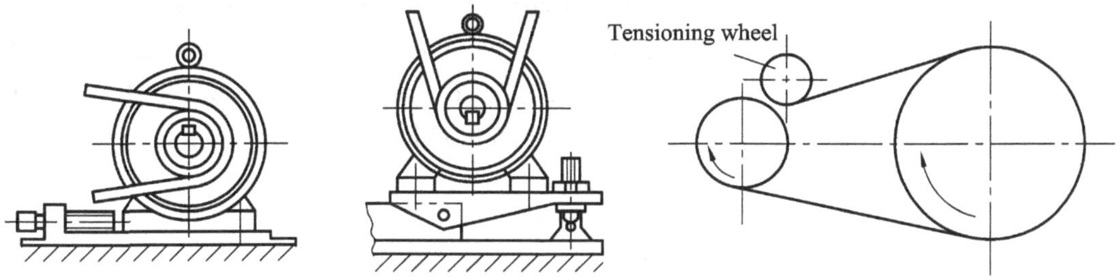

(a) Method of changing the center distance (b) Method of using the tensioning wheel

Fig. 3-25 Adjustment methods for tension

V. Assembly of belt pulley and shaft

Generally, the connection between the belt pulley and the shaft is transition fit (H7 / k6). In order to transmit larger torque, the key and fastener are needed for circumferential fixation and axial fixation. Several ways of connection between the belt pulley and the shaft are shown in Fig. 3-26.

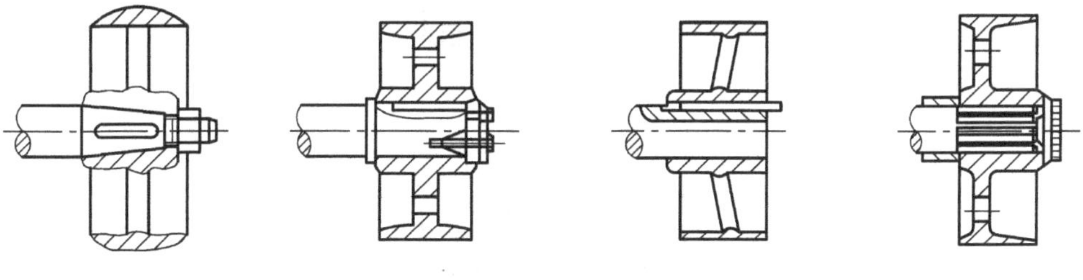

(a) For tapered journal, they are fixed with nuts (b) For cylindrical journal, shaft shoulder and retaining ring are fixed with screws (c) For cylindrical journal, they are linked with taper key (d) For cylindrical journal, the spacer sleeve, spline and retaining ring are fixed with screws

Fig. 3-26 Assembly of belt pulley and shaft

When assembling, use a wooden hammer a red copper rod to drive the belt pulley into the shaft, or use screw press to press it into the shaft, as shown in Fig. 3-27.

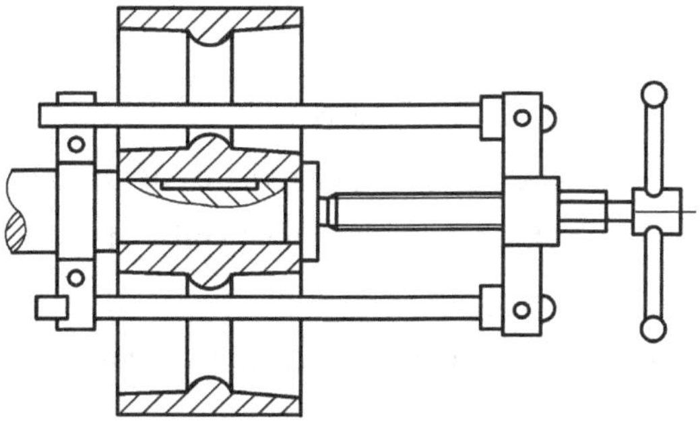

Fig. 3-27 Screw press tool

When the belt pulley is assembled, it is necessary to check the radial circular runout and end face circular runout of the belt pulley with a marking block or dial indicator. The inspection method is shown in Fig. 3-28. In addition, it is necessary to ensure the correctness of their mutual positions, so as to prevent uneven belt tension caused by the inclination or dislocation of the belt pulley, which may lead to excessive wear of the belt. The inspection method is shown in Fig. 3-29. When the center distance is large, the pulling method is used, and when the center distance is small, the ruler can be used for measurement.

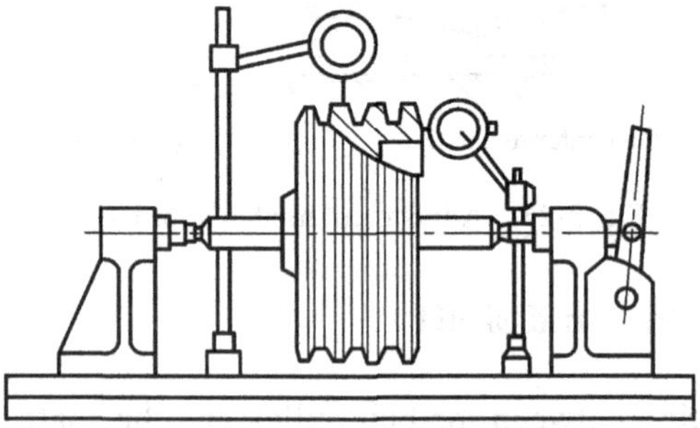

Fig. 3-28 Inspection of belt pulley runout

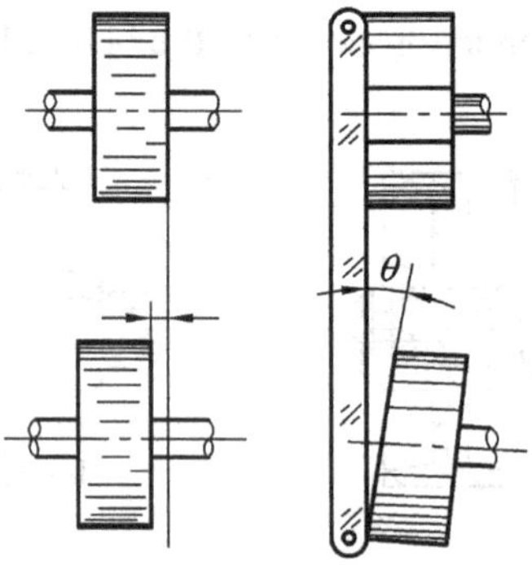

Fig. 3-29 Inspection of the correctness of the mutual position of the belt pulley

Task Three Assembly of chain drive mechanism

※ Knowledge objectives and skill objectives

(1) Understand the classification and characteristics of chain drive mechanism.

(2) Master the assembly method and adjustment skills of chain drive.

I. Types and characteristics of chain drive mechanism

1. Types of chain drive mechanism

According to different purposes, chain drive mechanism can be divided into the following three types.

(1) Transmission chain. It is mainly used to transmit motion and power in general machinery, and can also be used in transportation and other occasions.

(2) Conveyor chain. It is used to transport workpieces, articles and materials. It can also be directly used in various machines and can also form a chain conveyor as a unit.

(3) Traction chain. It is mainly used to transmit force, play the role of traction, suspension, and also do slow motion.

Since the transmission chain has the widest application range, only the transmission chain is introduced here. The transmission chain mainly has the following forms: sleeve chain, sleeve roller chain, toothed chain, as shown in Fig. 3-30 and Fig. 3-31.

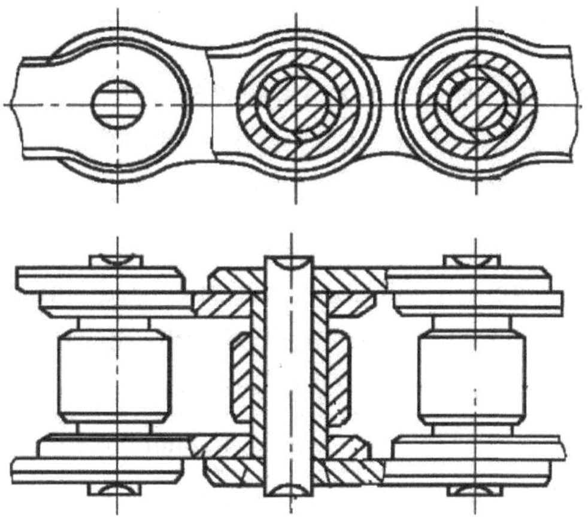

Fig. 3-30 Sleeve roller chain

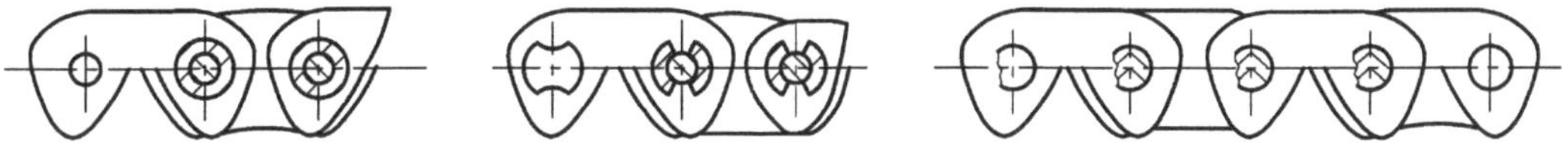

(a) (b) (c)

Fig. 3-31 Toothed chain

2. Transmission characteristics of chain drive mechanism

Chain drive has the characteristics of both gear drive and belt drive. It belongs to meshing drive with intermediate flexible parts. Compared with gear drive and belt drive, the main advantages of chain drive are: no elastic sliding, accurate average transmission ratio; higher transmission efficiency, smaller chain tension on the shaft; able to work under harsh conditions; large bearing capacity; low requirements for manufacturing and installation accuracy; large shaft spacing and simple structure. The shortcomings of the chain drive are: poor transmission stability and noise; the instantaneous transmission ratio is not constant; pone to tooth skipping and chain loss after wear; poor performance of rapid reverse rotation; not suitable for the drive with too small shaft spacing or rapid reversing.

II. Assembly of chain drive mechanism

1. Technical requirements for assembly of chain drive mechanism

(1) The two axes of the sprocket must be parallel. Non-parallel will aggravate the wear of chain and sprocket, reduce the transmission stability and increase the noise. The inspection method is shown in Fig. 3-32, and the error is determined by measuring dimensions *A* and *B*.

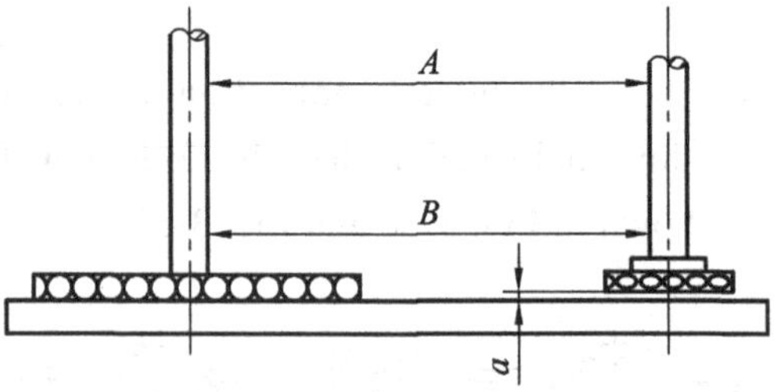

Fig. 3-32 Inspection of parallelism and axial offset of the two axes of the sprocket

(2) The axial offset of the two sprockets must be within the required range. Generally, when the center distance is less than 500 mm, the allowable offset *a* is 1 mm; when the center distance is more than 500 mm, the allowable offset *a* is 2 mm.

(3) The runout of sprocket should meet the specified value. The allowable runout of sprocket is shown in Table 3-1, which can be checked with a marking block or a dial indicator.

Table 3-1 Allowable runout of the sprocket mm

Sprocket diameter	Sprocket runout of sleeve roller chain	
	Radial circular runout	End face circular runout
≤100	0.25	0.3
100-200	0.5	0.5
200-300	0.75	0.8
300-400	1.0	1.0
>400	1.2	1.5

(4) The sag of the chain should be proper. If the sag is too tight, it will aggravate the wear; if the sag is too loose, it is easy to produce off-chain or vibration. The inspection method is shown in Fig. 3-33. In general, the sag *f* should not be greater than 20% *L* in horizontal transmission; in vertical position, *f* should not be greater than 0.2% *L*, and *L* is the center distance between the two chains.

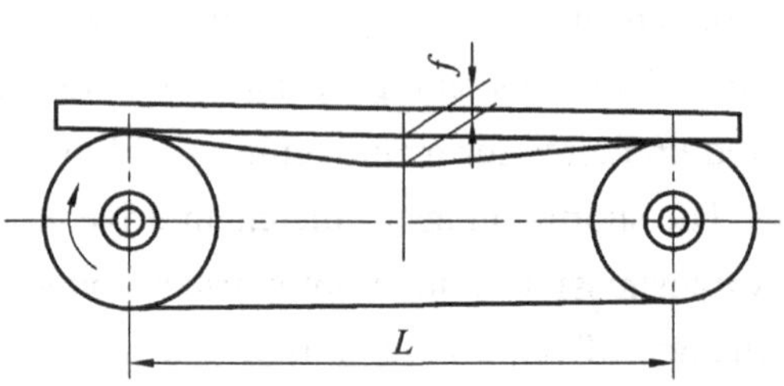

Fig. 3-33 Inspection method of chain sag

2. Assembly method of chain drive mechanism

The fixing method of sprocket on shaft is shown in Fig. 3-34. Fig. 3-34 (a) shows that the chain wheel is connected with a key and then fixed with a set screw. Fig. 3-34 (b) shows that the chain wheel is fixed with a tapered pin. The assembly method of the chain wheel is basically the same as that of the belt pulley.

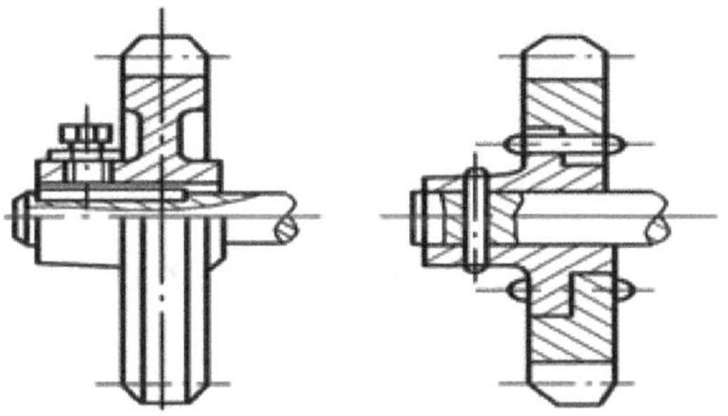

(a) (b)

Fig. 3-34 Fixing method of sprocket

The joint form of the sleeve roller chain is shown in Fig. 3-35. Fig. 3-35 (a) shows the movable pin fixed with a split pin; Fig. 3-35 (b) shows the movable pin fixed with a spring card. Both of them are applicable when the number of chain lines is even. When using the spring card, the direction of the open end must be opposite to the speed direction of the chain, so as to avoid falling off due to collision during operation. Fig. 3-35 (c) shows the case of using transitional link joints, which is applicable when the number of chain lines is odd. This kind of transition line has good flexibility and has the function of buffering and vibration absorption, but this kind of line plate will be subject to additional bending effect, so the odd number of chain lines should be avoided as far as possible.

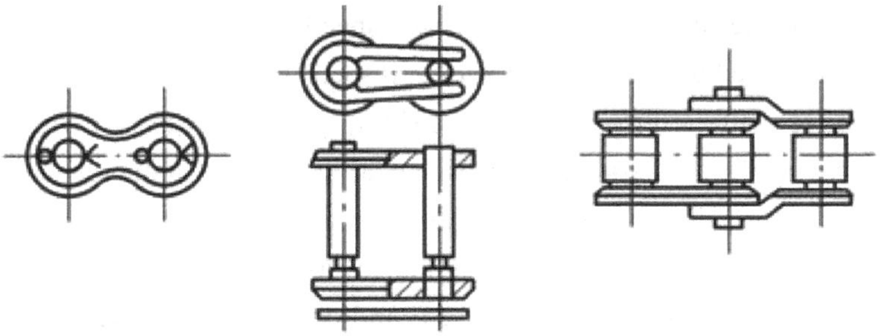

(a) (b) (c)

Fig. 3-35 Joint form of the sleeve roller chain

For the joining of both ends of the chain, if the center distance between the two shafts is adjustable and the sprocket is at the shaft end, it can be connected in advance and then installed on the sprocket. If the structure does not allow the chain to be connected in advance, the chain must be put on the sprocket first, and then be tightened with a special tensioning tool. Fig. 3-36 shows tools for tightening the chain.

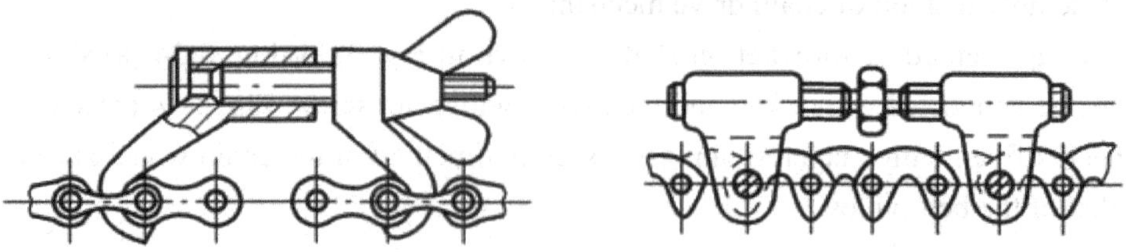

(a) (b)

Fig. 3-36 Tools for tightening the chain

Task Four Assembly of gear drive mechanism

※ Knowledge objectives and skill objectives

(1) Understand the classification and transmission characteristics of gear drive.

(2) Master the calculation method of basic parameters and geometric dimensions of straight spur gear.

(3) Master the assembly and inspection of straight spur gear drive mechanism.

I. Classification and characteristics of gear drive

1. Classification of gear drive

There are many kinds of gear drive. They can be classified according to different methods.

(1) According to the different circumferential speed, gear drive can be divided into three types: low speed drive, medium speed drive and high speed drive.

(2) According to the relative position of the shaft, gear drive can be divided into two types: plane gear drive (two shafts are parallel) and space gear drive (two shafts are not parallel).

(3) According to the different working conditions, gear drive can be divided into three types: closed gear drive (enclosed in the box and good lubrication ensured), semi-open gear drive (gear immersed in oil pool, with shield, but not closed) and open gear drive (gear exposed and good lubrication can not be ensured).

(4) According to the meshing mode of gear, gear drive can be divided into three types: external meshing gear drive, internal meshing gear drive and rack gear drive.

2. Characteristics of gear drive

(1) The range of power and speed transmitted by gear drive is large, and the circumferential speed can reach 300 m/s.

(2) Gear drive can ensure constant instantaneous transmission ratio, high transmission

stability, accurate and reliable transmission movement.

(3) The structure of gear drive is compact, which can help realize larger transmission ratio.

(4) Gear drive has high transmission efficiency and long service life.

(5) The manufacturing and installation accuracy requirement of gear drive is high, so the cost is also high.

II. The calculation method of basic parameters and geometric dimensions of straight spur gear

1. Basic parameters of straight spur gear

The basic parameters of straight spur gear mainly include the number of teeth z, modulus m, nominal pressure angle α, addendum coefficient h_a * and bottom clearance coefficient c *.

(1) Number of teeth z: The total number of teeth of a gear is the number of teeth. When there are certain modulus of the gear, the more the number of teeth, the larger the geometric size of the gear, the larger the curvature radius of the tooth involute, and the straighter the tooth profile curve.

(2) Modulus m: The quotient obtained by dividing the tooth pitch p by the circumference ratio π is the modulus, that is, $m=P/\pi$, and the unit is millimeter. For gears with the same number of teeth, the larger the module, the larger the size, the larger the teeth, and the stronger the bearing capacity; for gears with the same reference circle diameter, the larger the module, the stronger the bearing capacity, but the smaller the number of teeth. The standard modulus table is shown in Table 3-2.

Table 3-2 Standard modulus

The first series	0.1	0.12	0.15	0.2	0.25	0.3	0.4	0.5	0.6	0.8	
	1	1.25	1.5	2	2.5	3	4	5	6	8	
	10	12	16	20	25	32	40	50			
The second series	0.35	0.7	0.9	1.75	2.25	2.75	(3.25)	3.5	(3.75)	4.5	5.5
	(6.5)	7	9	(11)	14	18	22	28	(30)	36	45

Note: When selecting modulus, the first series should enjoy priority, followed by the second series, and the modulus in brackets should not be used as much as possible.

(3) Nominal pressure angle α: On the end plane, the acute angle between the radial straight line passing through any point on the tooth profile of the end point surface and the tangent line of the tooth profile at that point is the nominal pressure angle of that point. It is stipulated that $\alpha = 20°$.

(4) Addendum coefficient h_a*: The addendum is directly proportional to the modulus, that is, $h_a = m \cdot h_a$*. The addendum coefficient of standard straight spur gears h_a* = 1.

(5) Bottom clearance coefficient c^*: When the gears mesh, there should be a certain radial clearance, which is bottom clearance, expressed by c, and $c= c^* m$. It is stipulated that $c^* = 0.25$.

2. The calculation of the geometric dimensions of the straight spur gear

The involute straight spur gear with standard module m, nominal pressure angle $\alpha = 20°$, addendum coefficient $h_a^* = 1$, Bottom clearance coefficient $c^* = 0.25$, and end face tooth thickness s equals to end face slot width e is standard straight spur gear.

The calculation formula of geometric dimensions of the standard straight spur gear are shown in Table 3-3.

Table 3-3 Geometric dimensions calculation formula of standard straight spur gear

Name	**Code**	**Calculation formula**
Nominal pressure angle	α	20° for standard gears
Modulus	m	$m = P/\pi$
Tooth thickness	s	$s = P/2$
Tooth width	e	$e = P/2$
Pitch	P	$P = m\pi$
Base circle pitch	P_b	$P_b = P \cos \alpha$
Addendum	h_a	$h_a = h_a^* m = m$
Dedendum	h_f	$h_f = (h_a^* + c^*)m = 1.25\ m$
Tooth depth	h	$h = h_a + h_f = 2.25\ m$
Diameter of reference circle	d	$d = mz$
Diameter of addendum circle	d_a	$d_a = m(z + 2)$
Diameter of dedendum circle	d_f	$d_f = d - 2h_f = m(z - 2.5)$
Diameter of base circle	d_b	$d_b = d \cos \alpha$
Standard center distance	a	$a = m(z_1 + z_2)/2$

III. Assembly and inspection of cylindrical gear drive mechanism

1. Technical requirements for assembly of gear drive mechanism

(1) The fit of the gear hole and the shaft should meet the requirements of use.

(2) It must be ensured that the gears have accurate installation center distance and proper tooth side clearance.

(3) It must be ensured that the tooth surface has a certain contact area and the correct contact position.

(4) In the speed change mechanism, the accurate positioning of the gear should be ensured, and the dislocation should not exceed the specified value.

(5) For large gear with higher rotation speed, the dynamic balance check is usually carried out after it is installed on the shaft to avoid excessive vibration.

2. Assembly of cylindrical gear drive mechanism

1) Assembly between the gear and the shaft

There are three types of connection between the gear and the shaft: fixed connection, empty sleeve connection and sliding connection. Fixed connection mainly includes key connection, bolt flange connection and fixed riveting; sliding connection mainly adopts spline connection (sliding key connection can also be used when the transmission torque is small.)

(1) Generally, gears with empty sleeves or sliding on the shaft are clearance fits, and the accuracy after assembly depends on the machining accuracy of the parts. The assembly is more convenient, and the gears shall not shake on the shaft after assembly.

(2) Most of the gears fixed on the shaft are transitional fits (with a small amount of interference). When the interference is not large, it can be knocked and pressed by hand tools; when the interference is large, it can be pressed by press machine. The three installation errors shown in Fig. 3-37 should be avoided.

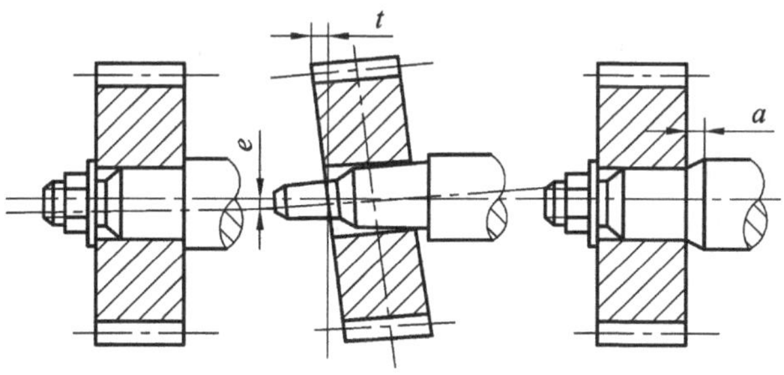

(a) (b) (c)

Fig. 3-37 Installation error of gear on shaft

(3) For the gear drive mechanism with high accuracy requirements, the radial circular runout error and end face radial circular runout error should be checked after press fitting.

Fig. 3-38 shows the detection method of gear radial circular runout error. Support the gear shaft on the V-shaped frame or two centers, make the shaft parallel to the plate, place the cylindrical gauge between the gear teeth, put the measuring head of the dial indicator on the cylindrical gauge, and read the number from the dial indicator. Then rotate the gear and repeat the measurement every 3-4 teeth. The difference between the maximum reading and the minimum readings of the dial indicator is the radial circle runout error on the gear reference circle.

Fig. 3-39 shows the detection method of gear end face circular runout error. Use the center to push the shaft in the middle, so that the measuring head of the dial indicator is against the end face of the gear. Within the range of one revolution of the gear shaft, the difference between the maximum reading and the minimum reading of the dial indicator is the end face circular runout error of the gear.

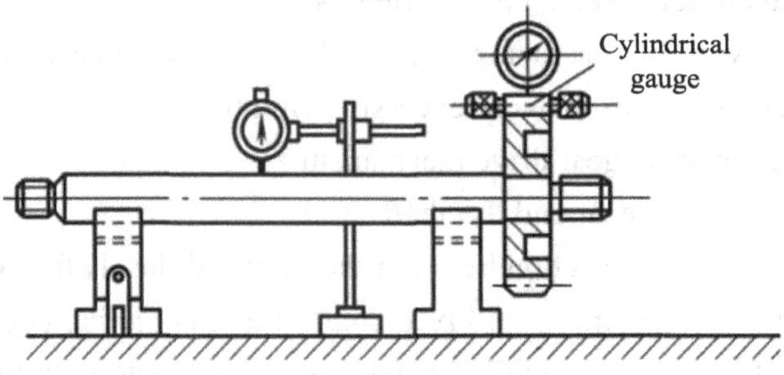

Fig. 3-38 Detection of the gear radial runout error

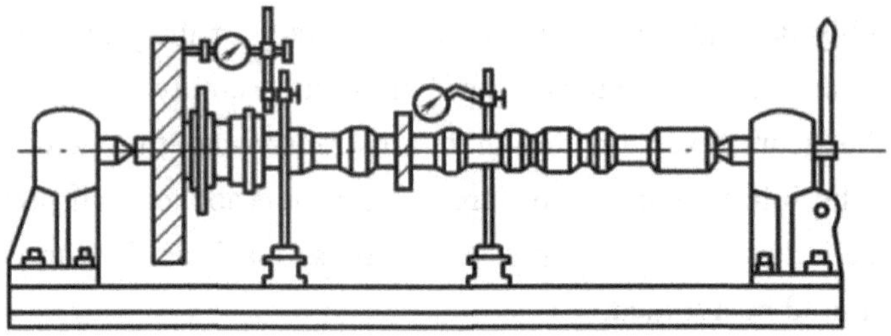

Fig. 3-39 Detection of the gear end face circular runout error

2) Install the gear shaft assembly pieces into the box

(1) Detection of the box hole accuracy before assembly.

◇ The detection of hole distance is shown in Fig. 3-40.

◇ The detection of hole parallelism is shown in Fig. 3-41.

◇ The detection of the perpendicularity between the hole end surface and the hole centerline is shown in Fig. 3-42.

◇ Detection of the coaxiality of the hole system as shown in Fig. 3-43.

◇ Check of the size accuracy and parallelism of hole center line and base surface.

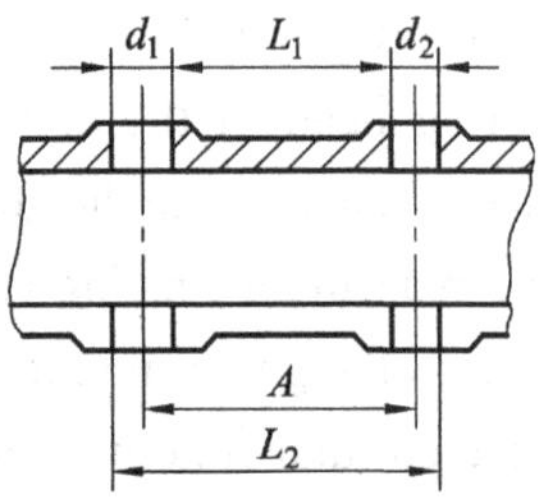

Fig. 3-40 Detection of hole distance

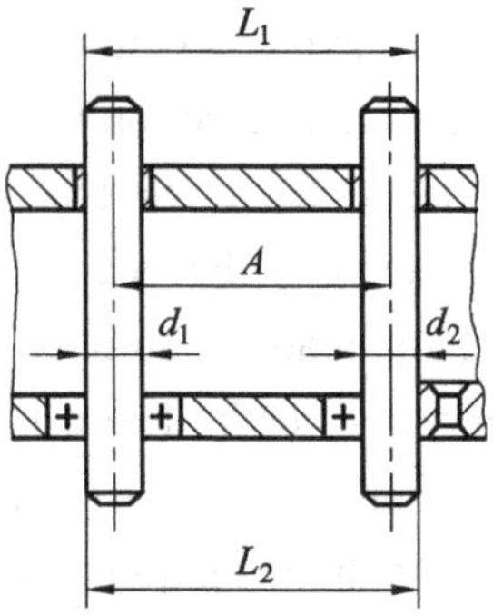

Fig. 3-41 Detection of hole parallelism

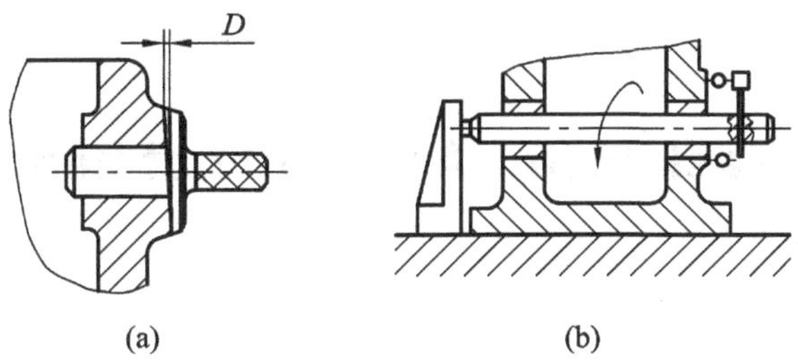

Fig. 3-42 Detection of the perpendicularity between the end face and the centerline of the hole

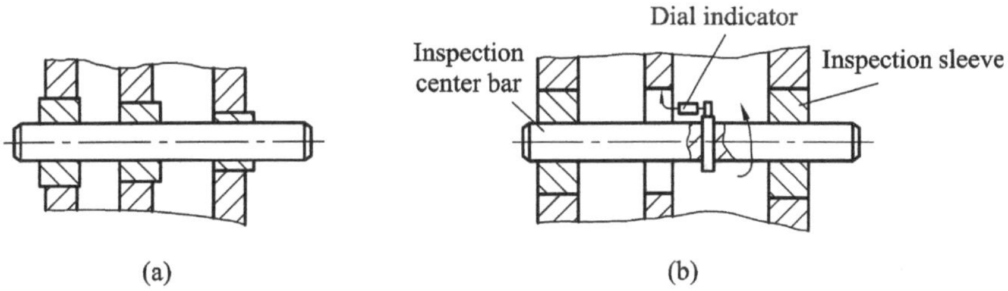

Fig. 3-43 Detection of the coaxiality

(2) Inspection of the meshing quality of gears.

Check the tooth side clearance with a pressure fuse, as shown in Fig. 3-44.

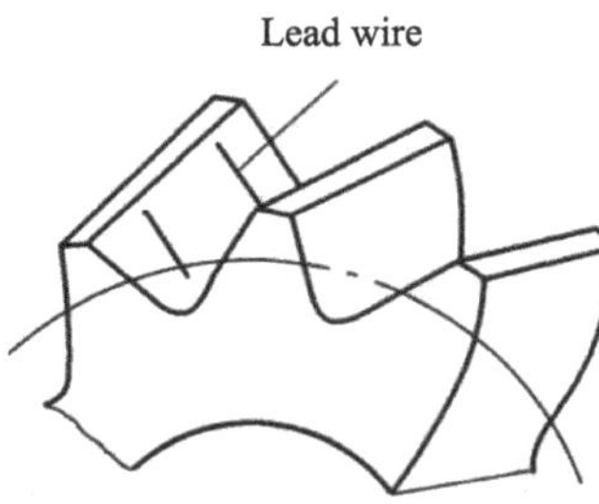

Fig. 3-44 Inspection of pressure fuse

Place (stick with grease) two fuse wires on the tooth surface along both ends of the tooth length and perpendicular to the tooth length direction (3 to 4 for wide teeth). The diameter of the fuse wire should not be more than 4 times of the minimum limit of side clearance (usually 1.25 to 1.5 times of the side clearance) specified by the gear pair. After the rolling gear is squeezed, the thickness of the thinnest part of the fuse wire is measured, which is the side clearance of the gear pair.

When checking the contact accuracy, the coloring method should be used. Apply red lead powder to the tooth surface of the driving gear. After turning the driving gear and braking the driven gear slightly, check the contact spots. For gears working in both directions, both forward and reversing directions should be checked.

Task Five Assembly of bearing

※ Knowledge objectives and skill objectives

(1) Understand the structure and code of rolling bearing, sliding bearing and shaft group.

(2) Master the selection method and assembly knowledge of bearing.

(3) Understand the types and characteristics of commonly used lubricants and greases.

I. Assembly of rolling bearing

The bearing is used to support the rotating shaft and reduce the friction and wear caused by the relative rotation between the rotating body and the supporting body. Bearing includes rolling bearing for rotating motion, sliding bearing for linear motion, and joint bearing for swinging or tilting motion.

(I) Classification and code of rolling bearing

Rolling bearing is standard components manufactured by manufacturers according to the classification method of bearing. There are many types of rolling bearing. They have complex models and different specifications. Radial and angular contact bearing is generally composed of inner ring, outer ring, rolling element and cage. Thrust bearing is composed of shaft ring, rolling element, cage and seat ring, as shown in Fig. 3-45.

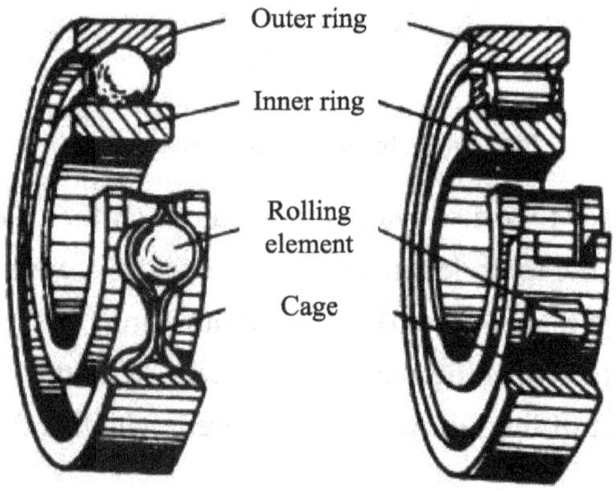

Fig. 3-45 The structure diagram of the rolling bearing

1. Classification of rolling bearing

Rolling bearing can be divided into different categories according to the direction of load that can be borne or the nominal contact angle:

(1) Radial bearing. It is mainly used for bearing radial load, and its nominal contact angle ranges from 0° to 45°.

(2) Thrust bearing. It is mainly used for bearing axial load, and its nominal contact angle ranges from 45° (45° not included) to 90°.

Bearing can be divided into several types according to their rolling elements:

(1) Ball bearing. The rolling elements are balls.

(2) Roller bearing. The rolling elements are rollers. According to the types of rollers, roller bearing divided into the following types:

◇ Cylindrical roller bearing. The rolling elements are cylindrical rollers. The ratio of the length to the diameter of the cylindrical roller is less than or equal to 3.

◇ Needle roller bearing. The rolling elements are needle rollers. The ratio of needle length to diameter is greater than 3, but the diameter is less than or equal to 5mm.

◇ Tapered roller bearing. The rolling elements are tapered rollers.

◇ Spherical roller bearing. The rolling elements are spherical rollers.

Bearing can also be divided the following types according to the number of rows of rolling elements:

(1) Single row bearing with one row of rolling elements.

(2) Double row bearing with two rows of rolling elements.

(3) Multi-row bearing with more than two rows of rolling elements, such as three row and four row bearing.

Bearing can be divided into multiple structural types according to their structural shapes (such as with or without filling slot, with or without inner or outer ring as well as the shape of rings, the structure of the ribs, and even with or without cages, etc.).

2. Code of rolling bearing

There are many types of rolling bearing. For the convenience of selection, the national standard stipulates that the rolling bearing shall be represented by a code. The code can indicate the structure, size, tolerance level and technical performance of the rolling bearing. The code of rolling bearing is composed of letters and numbers. The complete code includes three parts: pre-code, basic code and post-code, as shown in Table 3-4.

Table 3-4 Code of rolling bearing

Pre-code	Basic code					Post-code
	Number or letter	Number	Number	Number	Number	
	Type	Width (height) series code	Diameter series code	Inner diameter of the Bearing		

1) Pre-code

The Pre-code of the rolling bearing is the supplementary code added in front of the code when the shape, size, tolerance or other technical requirements of the bearing are changed. The pre-code of the rolling bearing is indicated by letters and the meaning of the letters is shown in Table 3-5.

Table 3-5 The meaning of the letters

Code	Meaning	Example
L	Separable inner ring or outer ring of separable bearing	LNU205
R	Bearing without separable inner or outer rings (needle roller bearings are only available for NA type)	RNU205
K	Roller and cage assembly	K81105
WS	Thrust cylindrical roller bearing washer	WS81105
GS	Thrust cylindrical roller bearing race	GS81105
F	Radial ball bearing with flanged outer ring (only for $d \leqslant 10$ mm)	F619/5
KOW	Thrust bearing without shaft ring	KOW-51105
KIW	Seatless thrust bearing	KIW-51106
LR	Bearing with separable inner or outer ring and rolling element assembly	—

2) Basic code

The basic code is composed of three parts: bearing type code, size series code and inner diameter code, arranged in order from left to right. The bearing type code is represented by Arabic numerals (hereinafter referred to as numbers) or uppercase Latin letters (hereinafter referred to as letters), and the size series code and inner diameter code are represented by numbers.

(1) The bearing type code is shown in Table 3-6.

Table 3-6 Bearing type code

Bearing type code	Bearing type	Bearing type code	Bearing type
0	Double row angular contact ball bearing	6	Deep groove ball bearing
1	Self-aligning ball bearing	7	Angular contact ball bearing
2	Spherical roller bearings and thrust spherical roller bearing	8	Thrust cylindrical roller bearing
3	Tapered roller bearing	N	Cylindrical roller bearing; double or multiple rows are indicated by the letter NN
4	Double row deep groove ball bearing	U	outer spherical ball bearing
5	Thrust ball bearing	QJ	Four-point contact ball bearing

(2) The size series code is composed of the width (or height) series code (one number) and the diameter series code (one number) of the bearing. It reflects the width and thickness of the inner and outer rings of the same bearing when the inner ring diameter is the same, and the size of the rolling elements. Obviously, bearing with different size series codes have different outer dimensions and different bearing capacities.

The size series code can sometimes be omitted. Except tapered roller bearing, the "0" representing the width series code of other types of bearings can be omitted; the "1" in the 10 size series codes of deep groove ball bearing and angular contact ball bearing can be omitted; The width series code "2" of the double row deep groove ball bearing can be omitted. The size series code of radial bearing and thrust bearing are as shown in Table 3-7, Table 3-8, and Table 3-9.

Table 3-7 Size series code of radial bearing and thrust bearing

Diameter series code	Radial bearing								Thrust bearing			
	Width series code								Height series code			
	8	0	1	2	3	4	5	6	7	9	1	2
	Size series code											
7	—	—	17	—	37	—	—	—	—	—	—	—
8	—	08	18	28	38	48	58	68	—	—	—	—
9	—	09	19	29	39	49	59	69	—	—	—	—
0	—	00	10	20	30	40	50	60	70	90	10	—
1	—	01	11	21	31	41	51	61	71	91	11	—
2	82	02	12	22	32	42	52	62	72	92	12	22
3	83	03	13	23	33	—	—	—	73	93	13	23
4	—	04	—	24	—	—	—	—	74	94	14	24
5	—	—	—	—	—	—	—	—	—	95	—	—

Table 3-8 Meanings of width series code and Height series code

Code	**Width serices code**								**Height series code**			
	8	**0**	**1**	**2**	**3**	**4**	**5**	**6**	**7**	**9**	**1**	**2**
Meaning	Extra narrow	Narrow	Normal	Wide	Extra wide 3	Extra wide 4	Extra wide 5	Extra wide 6	Extra low	Low	Normal	Wide

Table 3-9 Meanings of diameter series code

Code	**7**	**8**	**9**	**0**	**1**	**2**	**3**	**4**	**5**
Meaning	Super Extra light	Super light	Super light	Extra light	Extra light	Light	Medium	Heavy	Extra heavy

(3) Code of inner diameter. The inner diameter of the rolling bearing is represented by

two-digit numbers. The specific way to indicate the inner diameter of the bearing is shown in Table 3-10.

Table 3-10 The inner diameter indicator of the rolling bearing

Nominal inner diameter of bearing /mm		Inner diameter code	Example
0.6–10(non-integer)		Directly expressed in millimeters of the nominal inner diameter, and is separated from the dimension series code by "/"	Deep groove ball bearing 618/2.5 $d = 2.5$ mm
1–9(integer)		Directly expressed in millimeters of the nominal inner diameter. For deep groove and angular contact ball bearings 7,8,9 diameter series, the inner diameter and the dimension series code are separated by "/"	Deep groove ball bearing 625 618/5 $d = 5$ mm
10–17	10	00	Deep groove ball bearing 6200 $d = 10$ mm
	12	01	
	15	02	
	17	03	
20–480(except 22,28,32)		For the quotient of the nominal inner diameter divided by 5, when the quotient is a single digit, then "0" must be added to the left of the quotient, such as 08	Spherical roller bearing 23208 $d = 40$ mm
Greater than and equal to 500, and 22,28,32		Directly expressed in millimeters of the nominal inner diameter, but separated from the dimension series by "/"	Spherical roller bearing 230/500 $d = 500$ mm Deep groove ball bearing 62/22 $d = 22$ mm

3) Post-code

The post-code is represented by letters (or followed by numbers), and placed on the right side of the basic code and half a Chinese character distance from the basic code (The situations are exceptional when there are symbols "-" or "/" in the code). When there are many changes and multiple groups of post-codes, they are arranged in order from left to right as shown in Table 3-11.

Table 3-11 The composition of the post-code

Code	1	2	3	4	5	6	7	8
Meaning	Internal structure	Seal and dust ring modification	Cage and its material	Bearing material	Tolerance class	Clearance	Config uration	Others

(II) Selection principle of rolling bearing

1. Capacity of bearing

The size, direction and nature of the load on the bearing are the main basis for selecting the bearing type, and the limiting speed should also be considered. If the speed is high, the load is small, and the rotation accuracy is required to be high, ball bearing should be used; when the speed is low, the load is large or there is an impact load, then roller bearing should be used; when the radial load is the mainstay, deep groove ball bearing should be selected; when there is only radial load, low speed, large load or impact, cylindrical roller bearing can be used; when there is only axial force, thrust ball bearing can be used; when bearing both radial and axial loads, angular contact ball bearing and tapered roller bearing can be used. When the inner diameter is the same, the smaller the outer diameter, the higher the limit speed, so the light type is higher than the heavy type.

2. Alignment

When the two bearing housing holes have different axes or the bearing deforms greatly after loading, the inner and outer rings of the bearing will be relatively deviated, the self-aligning bearing should be used.

3. Installation and removal

Separate bearing should be used for bearing that is frequently assembled or disassembled.

4. Economy

Generally, ball bearing is cheaper than roller bearing. Care must be taken when selecting high-precision bearing. Only when the high-speed shaft have strict requirements for rotation precision, the high-precision bearing should be selected. For general mechanical transmission, bearing of ordinary precision can be used.

(III) Assembly of rolling bearing

Rolling bearing is standard part produced by professional factories, and their inner bore and outer diameter have been determined before leaving the factory. Therefore, the matching between the inner diameter of the bearing and the shaft is a base hole system, and the matching between the outer diameter and the housing hole is a base shaft system.

1. Assembly technical requirements for rolling bearing

(1) Before assembling, use kerosene to clean the bearing and remove the burr, rust and other defects on the mating surface.

(2) When assembling, the end face with the marking code should be installed in a visible direction so that it can be checked during replacement.

(3) The bearing must be close to the shaft shoulder or hole shoulder, and no gap or skew is allowed.

(4) For the two coaxial bearings, one of them must have space for axial movement when the shaft is heated or expanded.

(5) When assembling, the force should be acted evenly on the bearing ring to be fitted, and

it is not allowed to transmit pressure through the rolling elements.

(6) Keep clean during assembly and prevent foreign matter from entering the bearing.

(7) The assembled bearing should run flexibly, with low noise, and the temperature rise should not exceed the allowable value.

(8) The machining precision of the part matching the bearing should correspond to the precision of the bearing. Generally, the machining precision of the shaft shall be the same or higher precision of the bearing. The bearing seat hole shall be the same or has lower precision.

2. The fitting of rolling bearing

(1) The ring and shaft or housing hole that rotates relative to the load direction, should choose transition or interference fit. The size of the interference fit is based on the principle that when the bearing is working under load, the ring does not "crawl" on the fitting surface of the shaft or the housing hole.

(2) Generally speaking, heavy-load bearing should be installed tighter than light-load bearing and normal-load bearing. The heavier the load, the greater the coordination interference.

(3) The tolerance grade of fitting parts such as shaft or housing hole is related to bearing precision.

(4) In most cases, the rotation of the shaft and the radial load direction does not change, that is, when the inner ring of the bearing rotates relative to the load direction, the transition or interference fit should be selected generally.

3. Assembly method of rolling bearing

When assembling the bearing, the most basic requirement is to make the applied axial force directly act on the end face of the outer ring (when installed on the shaft, the applied axial force should directly act on the inner ring; when installed in the hole, the force should act directly on the outer ring). Try not to affect the rolling element. The assembly method include hammering method, pressing machine assembly method, hot assembly method and so on.

(1) Hammering method

The hammering method refers to the method of hammering and assembling with a hammer padded with a copper rod and some soft materials. It should be minded that do not let foreign matter such as copper powder fall into the bearing raceway, and do not directly hit the inner and outer rings of the bearing with a hammer or punch, so as not to affect the matching precision of the bearing or cause damage to the bearing.

(2) Pressing machine assembly method

For bearing with larger interference tolerances, screw pressing machines or hydraulic pressing machines can be used for assembly. Before pressing, the shaft and the bearing should be leveled, and a little lubricating oil should be applied. The pressing speed should not be too fast. After the bearing is in place, the pressure should be removed quickly to prevent damage to the bearing or shaft.

(3) Hot assembly method

The hot assembly method is to heat the bearing in oil to 80–100°C, so that the inner hole of the bearing is expanded and then set on the shaft, which can prevent the shaft and the bearing from being damaged. For the bearing which has dust caps and seals, and been filled with grease, hot assembly method is not applicable.

II. Assembly of sliding bearing

Sliding bearing has the advantages of simple structure, convenient manufacturing, small radial size, lubricating oil film adsorption, and ability to withstand large impact loads, as shown in Fig. 3-46. The sliding bearing works smoothly, reliably and without noise. Under liquid lubrication conditions, the sliding surface is separated by lubricating to avoid direct contact, which could greatly reduce friction and surface wear. The oil film also has the capability of vibration absorption to some extent, but the sliding bearing has a relatively large starting frictional resistance. The part of the shaft supported by the bearing is the journal, and the part matching the journal is the bearing bush. In order to improve the friction properties of the bearing pad surface, the anti-friction material layer cast on the inner surface is the bearing liner. The materials of bearing bushes and bearing linings are collectively referred to as sliding bearing materials. Sliding bearing is generally used under low-speed and heavy-load conditions or in operating parts where maintenance and lubrication are difficult to operate.

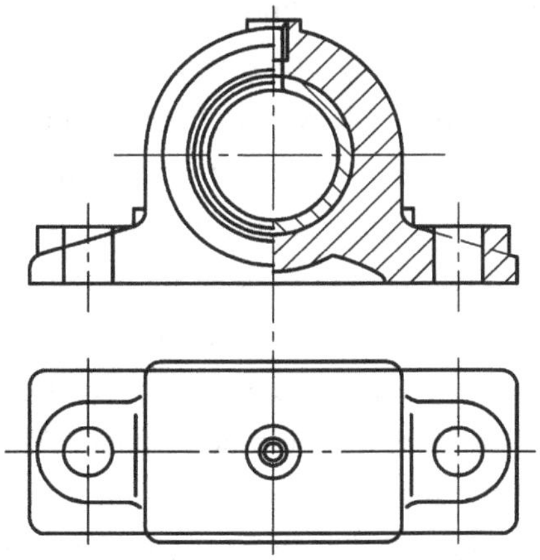

Fig. 3-46 Integral radial sliding bearing

(I) Classification of sliding bearing

1. Code of the sliding bearing

The code system of sliding bearing consists of Latin letters, Greek letters, Arabic numbers and other symbols. The simplest sliding bearing code is represented by a capital letter or lowercase letter; in special cases, it is composed of two or three capital letters. The most complicated code is the basic code with a subscript.

2. Classification of sliding bearing

(1) According to the direction of load bearing, sliding bearing can be divided into radial (centripetal) sliding bearing and thrust (axial) sliding bearing.

(2) According to the types of lubricants, sliding bearing can be divided into the following seven types: oil lubricated bearing, grease lubricated bearing, water lubricated bearing, gas bearing, solid lubricated bearing, magnetic fluid bearing and electromagnetic bearing.

(3) According to the thickness of lubricating film, sliding bearing can be divided into thin film lubricated bearing and thick film lubricated bearing.

(4) According to the bearing bush material, the sliding bearing can be divided into bronze bearing, cast iron bearing, plastic bearing, gemstone bearing, powder metallurgy bearing, self-lubricating bearing and oil bearing.

(5) According to the bearing bush structure, sliding bearing can be divided into circular bearing, elliptical bearing, three blade bearing, stepped surface bearing, tilting pad bearing and foil bearing.

(II) Assembly of sliding bearing

The sliding bearing is a kind of sliding friction bearing, which is characterized by stable and reliable operation, low noise, and it can withstand heavy load and relatively large impact load. According to the different structural forms, it can be divided into integral sliding bearing, split sliding bearing and tile sliding bearing.

1. Assembly of integral sliding bearing

The integral sliding bearing, commonly known as sleeve, is also the simplest form of the sliding bearing. It is mainly assembled by pressing and hammering. In special occasions, hot assembly method is adopted. Most of the sleeves are made of copper or cast iron and they should be assembled with special care. It can be assembled by hammering with a wooden hammer or a hammer with a wooden pad. When the interference tolerance is large, it can be pressed in with a pressing machine. Whether it is knocked in or pressed in, tilting must be avoided. After assembly, the oil groove and oil hole should be in the required position.

If the bearing is deformed after assembly, the inner hole should be trimmed. The smaller one can be cut by reaming, and the larger one should be scraped. At the same time, pay attention to make sure the fit clearance with the shaft within the tolerance range. In order to prevent the shaft sleeve from rotating during operation, the contact surface between the shaft sleeve and the box body should be equipped with positioning pins or straddle screws. Due to the different hardness of the materials of the box body and the shaft sleeve, it is easy to make the drill bit lean to the side of the soft material when drilling. The following solutions can be used: punch the side of the hard material with a sample punch before drilling, and then use the short drill bit to increase the rigidity of the drill bit when drilling.

2. Assembly of split sliding bearing

The split sliding bearing has the characteristics of simple structure, easy adjustment, convenient assembly and disassembly. There are two bearing bushes on the bearing body, and the reasonable clearance is adjusted by using gasket at the joint.

(1) Assembly of bearing bush and bearing body. When the bearing bush is assembled with

the bearing body, the upper and lower bearing bush must be in good contact with the inner hole of the bearing. If it does not meet the requirements, take the inner hole of the bearing of the thick wall bearing bush as the datum, scrape the back of the bearing bush, and make the steps at both ends of the bearing bush close to the two ends of the bearing body. As long as the split surface of the bearing bush is about 0.1 mm higher than that of the bearing body, there is no need to scrape.

(2) When the bearing bush is installed in the bearing body, no displacement is allowed in the radial or axial direction. Usually, the steps at both ends of the bearing bush are used to move or stop the positioning or the positioning pin is used for positioning.

(3) Matching scraping of the bearing bush. Generally, the split bearing bush uses the matching shaft grinding points. Usually, the lower bearing bush is scraped first, and then the upper one. In order to improve efficiency, the upper bearing bush and the cover can not be installed when the lower bearing bushes being scraped. When the contact points of the lower bearing bush basically meet the requirements, the upper bearing bush and cover can be pressed tightly. When the upper bearing bush is being scraped, the contact points of the lower bearing bush can be further adjusted. As the scraping goes on, the tightness of the shaft can be adjusted by changing the thickness of the gasket. The scraping is completed when the shaft can rotate easily without obvious clearance and the contact point meets the requirements given the situation that the bearing cover is fastened.

(4) Measurement of bearing clearance. The size of the bearing clearance can be adjusted by the gasket at the center split surface, or by directly scraping the bearing bush. The lead pressure method is usually used to measure the bearing clearance. Take several pieces of lead wire with a larger diameter than the bearing clearance, place them on the the journal and the split surface, then tighten the nut to compress the split surface tightly, screw off the nut, remove the bearing cap, carefully take out the flattened lead wire, and measure the thickness of each piece with a micrometer. The bearing clearance can be known according to the average thickness of the lead wire. Generally, the bearing clearance should be 1.5 ‰–2.5 ‰ (mm) of the shaft diameter, and the smaller value should be taken when the diameter is larger. If the shaft diameter is 60 mm, the bearing clearance should be between 0.09 mm and 0.15 mm.

III. Lubrication of bearing

In the working process of rolling bearing and sliding bearing, the internal components of the bearing have different degrees of relative sliding, which leads to frictional heating and part wear. Therefore, the bearing must be lubricated reliably in the work. When choosing lubricants, the temperature, working load and rotational speed of bearings should be considered. The main purpose of lubrication bearing is to reduce friction heat, avoid high working temperature, reduce wear, prevent rust, and achieve the function of heat dissipation and sealing.

1. The oil cup dripping lubrication

The oil cup dripping lubrication is a kind of lubrication method in which the oil is dripped to the bearing through the oil saving port in the oil cup. The advantages of oil cup dripping lubrication are simple in structure, convenient in use and oil-saving. The amount of oil supply can be adjusted by the oil saving port. Generally, one drop of oil every 3–8 seconds is appropriate, because too much oil will increase the temperature rise of the bearing. The disadvantage of oil cup dripping lubrication is that there are certain requirements for the viscosity of the lubricating oil. The lubricating oil with high viscosity and no heat dissipation function cannot be used. The oil cup dripping lubrication is suitable for situations of low speed, light load and low working temperature.

2. Oil bath (immersion) lubrication

Oil bath lubrication is a very simple way of lubrication in which the bearing part is immersed in the lubricating oil and the oil is brought into other parts of the bearing through running. Oil bath lubrication is one of the most common lubrication methods. Considering the loss caused by the stirring and temperature rise during oil bath lubrication, for horizontal shaft, the height of the bearing part invading into the lubricating oil should be limited to a certain extent. Generally, the oil level should be controlled near the center of the lowest rolling part of the bearing. The oil bath lubrication is sufficient, but the oil supply is not easy to adjust. If there is no filter device in the oil tank, it is easy to bring impurities into the bearing and damage the bearing. Oil bath lubrication is generally suitable for low or medium speed applications, and it is more commonly used in low speed bearings.

3. Splash lubrication

Splash lubrication is a kind of lubrication method that oil is splashed into the bearing through other moving parts. The oil supply quantity of splash lubrication is not easy to adjust, and the lubricating oil level should not be too high, otherwise it is easy to result in stirring loss and temperature rise, and it is also easy to bring impurities in the oil tank into the bearing and damage the bearing. In splash lubrication, the oil splashes into the bearing through the rotating body (impeller or "oil throwing ring") installed on the shaft, and the bearing is not immersed in the oil.

4. Circulating oil lubrication

Circulating oil lubrication is a lubrication method in which the lubricating oil is pumped from the oil tank to the parts of the bearing to be lubricated by the oil pump, and then returned to the oil tank from the oil return port for reuse after filtration. Circulating oil lubrication includes oil injection lubrication, forced lubrication and spray lubrication. The circulating oil lubrication is sufficiently lubricated, the oil supply is easy to control, and the ability of heat dissipation and impurity removal is strong. Circulating oil lubrication has high reliability and it is suitable for situations of high speed, high temperature, heavy load and occasions with heat dissipation or impurity removal as the purpose. Circulating oil lubrication is a relatively ideal lubrication method, but it requires an independent oil supply system, and the manufacturing cost is relatively high. The oil supply system consists of oil pump, cooler, filter, oil tank and oil pipeline.

5. Oil mist lubrication

Oil mist lubrication is a kind of lubrication method in which the lubricating oil mist is obtained in the oil mist generator by mixing compressed air with a small amount of lubricating oil, and then transported to the parts of the bearing to be lubricated. The oil mist lubrication has the advantages of small stirring loss and temperature rise, oil saving and there are less impurities in the oil. Moreover, the air flow generated by the oil mist injection has a certain cooling effect, which can also effectively prevent the impurities from entering or discharging. Oil mist lubrication is suitable for high speed applications.

6. Oil injection lubrication

Oil injection lubrication is a kind of lubrication method that the lubricating oil is inject into the bearing to lubricate the bearing by high-pressure injection. When the bearing runs at a high speed, the rolling element and cage also rotate at a very high speed, thus forming an air vortex around the bearing. If the general lubrication method is adopted, the lubricating oil will not easily reach the parts to be lubricated. On the other hand, due to the centrifugal force generated by high-speed rotation, it is difficult for lubricating oil to remain in the lubricated area. In order to ensure that there is sufficient but not too much lubricating oil to supply the high-speed running bearing, the lubricating oil must be forced into the bearing from one end for lubrication, and then flow into the oil tank from the other end.

7. Oil air lubrication

Oil air lubrication is a lubrication method in which the system mixes a small amount of lubricating oil with gas under certain pressure, and then supplies it to the bearing. Oil-air lubrication is usually suitable for applications where the bearing runs at a high speed.

Questions for review

1. What are the commonly used tools for fixed connection and assembly?
2. What are the common types of thread?
3. What are the anti-loosening measures for threaded connection?
4. According to the structural characteristics and different applications, what types of the key connection can be divided into?
5. What types of the belt drive can be divided into?
6. What are the adjustment methods of the belt drive tension?
7. According to different applications, what kinds of the chain drive can be divided into?
8. What are the basic parameters of straight spur gear?
9. Briefly introduce the inspection method of gear contact precision.
10. Briefly introduce the types of rolling bearing.
11. Briefly introduce the meaning of rolling bearing code.
12. What are the forms of bearing lubrication?

Module Four Testing and debugging of mechanical equipment

Task One Testing of assembly quality

※ Knowledge objectives and skill objectives

(1) Understand the key points of equipment appearance inspection.

(2) Understand the methods, principles and contents of equipment precision inspection.

(3) Understand the lubrication, sealing and leakage prevention knowledge commonly used in mechanical equipment.

I. Inspection of appearance

(1) Check whether the machine is installed in the required position and whether the installation is stable. All anchor bolts shall meet the design requirements, and the exposed height of the anchor bolts after tightening should be the same.

(2) Check the wedge-shaped iron pad between the bottom of the machine base. The quantity and distribution position should meet the requirements of the technical documents.

(3) Check the surface of all kinds of pipes, and they should be smooth, clean, without deformation or cracks.

(4) Check the surfaces of measuring instruments, which should be free of coverings Monitor the dials of measuring instruments, and reading should be clear .

(5) Check the pipes of hydraulic oil, lubricating oil and cutting fluid. The threads shall be full without broken tooth in the middle. The inner wall of the pipe should be clean and the pipe mouth should be be free of burrs.

(6) Check whether the paint is intact, and wether there are any sign of detachment or collision.

(7) Check whether the valves are closed tightly.

(8) Check whether the movable parts of the machine are light and flexible when manually

turned.

(9) Check whether the electrical wiring is clear and tidy, and has a ground connection.

(10) Check the inspection report of pressure vessels, and they should be complete.

II. Precision inspection

The precision inspection in the assembly process is very important. If there is only assembly technology and lack of the correct inspection method, it is difficult to ensure that the product reaches the desired quality. Only the combination of proficient operating skills and correct inspection methods, and the use of more advanced and appropriate detection equipment can ensure the quality of the assembly.

1. Inspection methods and principles

In mechanical products, each part has its corresponding function and role. As for machine tool assembly, there are many items for the assembly precision inspection, but the emphasis should be placed on the mutual position precision and relative motion precision of the relevant parts matching, the shape of the fitting surface of the parts in the assembly, as well as the shape precision and size precision. In order to judge whether the assembly of parts meets the design requirements, the appropriate inspection methods can be selected for correct inspection according to the inspection principles.

(1) Inspection principles. According to the different factors such as structural characteristics, precision requirements of inspected parts and inspection equipment conditions, five inspection principles are specified for the inspection of shape and position error: the principle of comparing with the fitted elements, the principle of measuring coordinate value, the principle of measuring characteristic parameter, the principle of measurement run out and the principle of controlling time limit. In production, the specific inspection scheme can be formulated based on the inspection principles, according to the production situation of parts, assembly requirements and equipment conditions.

(2) Inspection methods. The shape and geometric position of the parts have different features and structures. Therefore, in the inspection process, the appropriate measurement method can be selected according to the specific elements of the parts to be tested. The common methods measurement inspection include direct measurement, indirect measurement, combined measurement or measuring gauge measurement. The inspection data can be used to judge whether the parts meet the requirements of design and assembly.

2. Inspection contents

The assembly precision is inspected one by one in the assembly or assembling of parts. The precision inspection can be divided into the following types:

(1) Precision inspection of basic parts. Taking machine tool equipment as an example, its basic parts include the machine body , column, beam, sliding seat and guide rail. These parts are the basis of each moving parts and also the main parts which directly affects the precision of

machine tool. The precision detection of the basic parts mainly include the straightness error of the guide rail, the parallelism error between the guide rails and the perpendicularity error between the guide rails.

(2) Inspection of the mutual position precision between the machine tool parts. After the basic parts are assembled, the assembly or overall assembly between the parts can only be carried out after the precision of each item is inspected and qualified. But after the assembly of parts, there will often be cumulative errors between assembly parts, which may even exceed the requirements of the overall assembly. Therefore, when assembling parts, it is necessary to specify the position precision between parts and the precision of parts which affect the overall assembly precision.

III. Lubrication, sealing and leakage prevention of mechanical equipment

1. The function and purpose of mechanical equipment lubrication

In the process of production, lubrication system plays a significant function. The normal operation of the lubrication system can directly affect the smooth production. Therefore, the quality of equipment lubrication is the critical condition for the smooth production. Equipment lubrication refers to using scientific management means in accordance with technical specifications to timely, correctly and appropriately lubricate equipment, while saving oil, and ensuring safe and normal operation of the equipment. At the same time, equipment lubrication is also an important part of equipment management and equipment maintenance. It is important to lubricate equipment well, keep equipment lubrication in a good condition and ensure the normal operation of the lubrication system, which are advantageous measures to ensure the normal operation of equipment, prevent accidents, reduce wear of parts, prolong the service life of equipment, reduce energy consumption and improve the production efficiency of equipment. It is also an important content of energy saving to apply lubricating materials reasonably and correctly to reduce the waste of lubricating materials.

2. Factors affecting equipment lubrication

(1) The quality of the lubricant. Due to different manufacturers and production processes, the quality of lubricants are also quite different, in other words, they are uneven in quality. It is not easy to tell the ones of good quality from the ones of poor quality. The conclusion must be drawn through scientific inspection, based on the understanding of the production process.

(2) Effective control of the lubricant deposits and contamination impurities. The long-term operation of mechanical equipment will produce metal particles, paint-like impurities (rust-proof paint on the inner surface of Lubricating Tank of mechanical equipment), muck, carbonaceous and smoky sediment in the lubricant. If the sediments and impurities in lubricant cannot be effectively controlled, they will cause fatal damage to the equipment. It is important to replace lubricant in a timely manner. Timely replacement of equipment lubricant can effectively control the deposits and impurities in the lubricant, thus improving the operation efficiency and

extending the life of the equipment. However, it is difficult to remove the deposits and impurities in the oil pipe and small diameter oil system. Even if the lubricant is completely replaced, it is still difficult to remove the impurities. At this time, only by stopping the equipment operation and cleaning the entire oil pipe system can the impurities be removed.

(3) The selection of the lubricant. Because of the different operating conditions of equipment (including load, temperature, speed, etc.), the different environment and medium in which the equipment lubrication parts are located, and the different performance requirements of lubricants, careful consideration should be given to the selection of lubricants.

3. Sealing and leakage prevention of mechanical equipment

The parts that can prevent leakage are called seals. The quality of sealing performance is an important indicator to measure the quality of machical equipment.

Leakage is one of the common failures of mechanical equipment. There are two main reasons for the leakage: the surface of mechanical products have various defects, shape and size deviation, therefore, there will be inevitably gaps at the connection of mechanical parts, which will lead to leakage; there is pressure difference on both sides of the seal, and the working medium will leak through the gap formed by the pressure difference.

Reducing or eliminating the gap is the main way to prevent leakage. The function of sealing is to seal the gap between the joint surfaces, increase the resistance in the leakage channel, isolate or cut off the leakage channel; or add small work elements to the channel, which will cause pressure on the leakage and partially or completely balance off the pressure difference leading to leakage to prevent leakage.

(1) Basic requirements for sealing. The basic requirements for sealing are tightness and reliability, compact structure, simple and easy manufacturing, convenient maintenance, and long service life. The requirements for its materials are good strength, high hardness, plasticity, elasticity, high temperature resistance, aging resistance, and wear resistance.

(2) Classification of sealing. Sealing can be divided into two categories: static sealing between relatively static joint surfaces and dynamic sealing between relatively moving joint surfaces. Static sealing mainly includes three types: point sealing, rubber sealing and contact sealing. According to the working pressure, static sealing can be divided into medium and low pressure static sealing and high pressure static sealing. Medium and low pressure static seals usually adopt gasket seals which are softer in material and wider in size, while high-pressure static seals are metal gaskets which are harder in material and narrow in size. Dynamic sealing can be divided into two basic types: rotary sealing and reciprocating sealing. According to the situation whether the seal is in contact with the relatively moving parts, dynamic sealing can be divided into contact sealing and non-contact sealing. Generally speaking, the contact sealing has good sealing property, but it is limited by friction and wear, and it is suitable for the occasions with low linear speed of sealing surface. The non-contact seal has poor sealing performance, which is suitable for high speed occasions.

Task Two Equipment installation and debugging

※Knowledge objectives and skill objectives

(1) Understand the main tasks of equipment installation.

(2) Understand the common methods of equipment installation.

(3) Understand the general specifications of equipment installation.

I. Preparation for installation

Before construction, the equipment and materials mentioned in the technical preparation sheet must be ready, and the mechanical equipment to be installed must be transported to the site in time. Unpack them for inspection, and do the disassembly, cleaning and lubrication.

(1) Equipment cleaning. After the equipment is unpacked for inspection and disassembled, the anti-rust agent, internal residual iron filings, rust, and dust accumulated during transportation and storage must be cleaned up before assembly.

The cleaning agents commonly used in installation include alkaline cleaning agent, cleaning agent containing non-ionic surfactant, petroleum solvent, anti-corrosion solvent for cleaning gas medium.

The anti-rust layer on the machined surface of the equipment shall not be removed with emery cloth or metal scraper, but only with clean cotton yarn, cotton cloth, wooden scraper or horn scraper. Dry oil can be cleaned with kerosene. The anti-rust paint can be cleaned with banana water, alcohol, turpentine or acetone. If the rust on the machined surface cannot be removed with oil, wipe it off with a cotton cloth dipped with acetic acid, and then removing the rust, wipe it with lime water to neutralize it, and finally wipe it dry with a clean cotton yarn or cloth. When the anti-rust dry oil applied to the gears in the gearbox is too thick, it can be cleaned with hot engine oil of 70–80°C or hot kerosene of 30–40°C. The hardened or hard-to-clean ball bearings can be washed with hot engine oil of 70–80°C, then washed with kerosene, and finally cleaned with gasoline. Do not use cotton yarn when cleaning the bearing, only use clean cotton cloth. If the cotton cloth is inconvenient to clean, use oil gun to inject kerosene or hot engine oil for cleaning.

(2) Equipment lubrication. Only after the internal and external parts of the equipment are cleaned, the oil can be added for lubrication. Lubricating oil must be tested and confirmed to meet the requirements before it be used. Before being applied to the equipment, the lubricating oil must be filtered, and the amount of oil filled should reach the specified oil mark position. all lubricating parts and oil holes should be filled with lubricating oil.

Lubricating grease is usually used in the situation where the seal is simple, it's not easy to be polluted and not necessary to be replaced frequently. Grease holds great practical lubrication meaning especially for high-speed motors, automatic devices and equipment that are not easy to refuel. At the same time, the lubricating grease is not easy to be affected by temperature, has a wide range of adaptation to the changes of load properties and movement speed, and is not easy to flow away on the vertical lubricating surface. A tough oil film can be formed when appropriate amount of graphite powder is added to the the grease, and it can slow down the motion in the reciprocating motion mechanism, avoid crawling and eliminate vibration. Due to the poor fluidity and low thermal conductivity, grease can not be used as circulating lubricant.

II. Installation precision of equipment

Equipment installation is a project that combines a series of equipment into a set of equipment or a production line to form a complete technical equipment system and ultimately form a production capacity. The equipment installation precision refers to the position precision between the individual equipment in order to ensure the correct linkage of the entire set of equipment in the installation project. A single equipment can achieve the required manufacturing precision through reasonable installation process and adjustment method.

Ⅲ. Factors affecting equipment installation precision

(1) The quality (precision) of the infrastructure. It including the geometric dimensions, location, elevation of different planes, flatness of upper plane, structural form, embedded depth, strength, settlement, inclination and seismic performance of infrastructure.

(2) Installation quality (precision) of the gasket iron and the anchor bolt. It including the quality of the gasket iron itself, contact quality of the gasket iron, the quality of the secondary grouting; position, elevation, perpendicularity, and tightening torque of anchor bolts.

(3) Selection of equipment measurement datum. The selection of equipment measurement datum is directly related to the quality of the entire equipment installation, alignment and leveling. During installation, the measurement datum is usually selected on the processing surface of the equipment base, body, shell, base, machine body, table plate and base plate.

(4) Assembly precision of bulk equipment. It including the relative motion precision between moving parts, the fitting precision and contact quality between mating surfaces, which will directly affect the operation quality of the equipment.

(5) Selection of measuring instruments. The precision of the measuring device must be consistent with the precision requirements of the measured device, otherwise it cannot meet the quality requirements.

(6) Influence of the equipment manufacturing quality. If the equipment manufacturing quality and the assembly quality in the equipment factory can not meet the precision requirements of manufacturing and assembly, the influence on the installation precision is the

most direct, and most of these quality problems cannot be solved on the construction site, and most of them need to be returned to the factory for processing, therefore, it is critical to inspect the quality carefully before the equipment leaves the factory.

(7) Influence of environment. For example, the temperature change has a great influence on the equipment foundation and the equipment itself. The installation of large and precision equipment is especially easy to be affected by the environment.

(8) The technical level of the operator and the error caused by the operation. The operation error is inevitable. It is crucial to control the operation error within the allowable range.

IV. Control methods of installation precision

The methods to improve the installation precision should start from the aspects of operator, machine, material, method and environment. In particular, it is necessary to emphasize the role of the operator. The operator with the corresponding technical level and sense of responsibility should be selected to engage in the corresponding installation work. The operator, coupled with appropriate and advanced processing technology, well-equipped and appropriate processing machinery, measuring instruments suitable for the installation precision, and appropriate environment, all of the above can contribute to the higher installation quality and precision.

(1) Try to eliminate and avoid the factors that affect the installation precision.

(2) According to the design precision and structural characteristics of the equipment, The operator should select the appropriate and reasonable assembly and adjustment methods. Make use of the position of the compensable parts or choose to install one or a group of suitable fixed compensable parts to balance off the excessive accumulated installation error.

(3) Choose reasonable inspection methods, including measuring instruments and methods, whose precision level should be consistent with the precision requirements of the equipment to be tested.

(4) If necessary, the repair method can be selected to supplement the compensation parts to offset the excessive accumulated installation error. This method is only used when the the existing problem cannot be solved by the adjustment method.

(5) Certain deviation is allowed in equipment installation, and the deviation and its direction should be determined reasonably. Some deviations are directional and generally specified in the technical documents of the equipment. when there are no provisions in the technical documents of the equipment, the following principles can be used:

◇ It is beneficial to offset the influence of the weight of equipment accessories after installation.

◇ It is beneficial to offset the influence of the force generated during the operation of the equipment.

◇ It is beneficial to offset the influence of parts' wear.

◇ It is beneficial to offset the influence of the oil film between friction surfaces.

IV. Equipment debugging and trial operation

The trial operation steps of mechanical equipment are: first no-load operation and then load operation; first single machine operation, then systematic operation, and finally linkage operation. The trial operation starts from the part, from the part to a set of parts, from a set of parts to a single (set) piece of equipment. For a set of linkage unit composed of several sets of equipment, every single equipment should be debugged separately and then systematically linked for trial operation.

Questions for review

1. What are the main contents of assembly quality inspection?
2. What are the main aspects of assembly precision inspection?
3. What is the function and purpose of equipment lubrication?
4. What are the basic requirements of mechanical equipment sealing?
5. Briefly introduce the factors that affect the precision of the equipment installation.
6. Brief introduce the control methods of the precision of the equipment installation.

Model test of fitter elementary theory

Model test of fitter elementary theory (1)

I. Tick the following statements right(√) or wrong(×)

1. When the two jaws of the vernier caliper fit together, the zero line of the main scale and the sliding vernier scale should be aligned. (   )

2. When the micrometer movable sleeve is rotated one circle, the micrometer screw will move 1 mm. (   )

3. Feeler gauge is also a limit gauge. (   )

4. When the bench vise clamps the workpiece, a long pipe can be put on the handle to increase the clamping force. (   )

5. The tool included angle is refers to the angle between the main and auxiliary cutting planes. (   )

6. The limit dimension refers to the two values between the size can range. (   )

7. The dimensional tolerance refers to sum of the high limit dimension and the low limit dimension. (   )

8. Only one marking datum needs to be selected for plane marking, and two marking datum should be selected for three-dimensional marking. (   )

9. The length of the saw blade is expressed by the center distance of the installation holes at both ends. (   )

10. After the saw blade is reversely installed, the sawing cannot be performed normally due to the change of the wedge angle. (   )

11. During filing, the pressure of both hands on the file should remain unchanged. (   )

12. The clearance angle on the main cutting edge of the drill is the largest at the outer edge, and it becomes smaller as it gets near the center. (   )

13. The main purpose of adding cutting fluid when drilling is to improve the surface quality of the hole. (   )

14. The cutting edge on the outer circle of the cylindrical countersink is the main cutting edge, which plays a main cutting role. (   )

15. When tapping the thread by motor, the calibrated part of the tap can not be completely

pulled out, otherwise it will cause thread damaging when it is withdrawn. (   )

16. According to different working principles, the anti-loosening device of the thread is divided into three categories: preventing loosening with additional friction, with mechanical method and other methods. (   )

17. Pins, which are used to determine the mutual position of parts, are usually called positioning pinned connection. (   )

18. The types of chain drive mechanisms can be divided into transmission chains, conveyor chains, and traction chain according to different uses. (   )

19. The characteristic of gear drive is the accuracy of the transmission movement. (   )

20. The purpose of mechanical device lubrication is to reduce friction and energy consumption, reduce surface wear, extend service life, and ensure normal operation. (   )

II. Multiple-choice questions

1. Vernier calipers can be divided into (   )mm, 0.05mm and 0.10mm according to the graduation value.

A. 0.02   B. 0.2   C. 0.1   D. 0.5

2. The feeler gauge is actually a (   ) gauge.

A. angle value   B. dimensions   C. limit   D. extremity

3. The surface that passes through the selected point of the cutting edge is tangent to the cutting edge and is perpendicular to the main movement direction is(   ).

A. front plane   B. cutting plane   C. base plane   D. orthogonal plane

4. Dimensional tolerance refers to (   ) of the high limit size and the low limit size.

A. sum   B. difference   C. product   D. quotient

5. The number of the marking datum selected for three-dimensional marking is (   )

A. one   B. two   C. three   D. four

6. When marking the processed surface, the (   ) coating is generally used.

A. white spray paint   B. chalk paint   C. blue oil

7. If the saw blade is reversely installed, its wedge angle becomes (   ).

A. unchanged   B. bigger   C. smaller

8. The saw blade has a kerf, and it can make the width of the saw slot on the workpiece (   ) the thickness of the back of the saw blade.

A. less than   B. equal to   C. greater than

9. When filing narrow and long planes and trimming dimensions, (   ) can be selected.

A. push file method   B. forward file method   C. cross file method

10. The selection of the section shape of the file depends on the (   ) of the workpiece.

A. shape of the file surface   B. size of the file surface

C. hardness of material   D. file surface accuracy

11. When drilling, the rotary motion of the drill bit around its axis is(   ).

A. feed movement B. main motion

C. rotation movement D. auxiliary movement

12. When the hole is to be drilled through, the feed speed must ( ).

A. decrease B. increase C. keep unchanged

13. Reaming processing belongs to the type of hole ( ).

A. rough machining B. semi-finishing machining C. finishing machining

14. Taper countersinks can be divided into four types: 60°, 75°, 90° and 120° , according to their taper angle, of which ( ) is the most used.

A. 60° B. 75° C. 90° D. 120°

15. When tapping blind hole threads, the depth of the bottom hole should be ( ) the required screw hole depth.

A. equal to B. greater than C. less than

16. The thread that rises from left to right is called ( ).

A. left-hand thread B. right-hand thread C. pipe thread D. sawtooth thread

17. In order to obtain reliable and fastened thread connection, the thread pair must be ensured to have certain ( ).

A. friction torque B. tightening torque C. pre-tightening force

18. The flat key can be divided into two types: ordinary flat key and ( ) flat key.

A. guide B. round head C. square head D. single round head

19. When assembling the V-belt drive mechanism, the middle planes of the two wheels are required to coincide, so the inclination angle should not exceed ( ).

A.10° B. 5° C. 0.1° D. 1°

20. In the same kind of parts, any assembly part can be installed into the parts without repair or adjustment, and the specified assembly requirements can be met. This assembly method is( ).

A. interchange method B. matching method

C. adjustment method D. repair method

Reference answer

I. Tick the following statements right(√) or wrong(×)

1-5 √ × √ × √ 6-10 × × × √ × 11-15 × × × × √ 16-20 √ √ √ × √

II. Multiple–choice questions

1-5 ACCAC 6-10 CACAA 11-15 BABCC 16-20 BCADA

Model test of fitter elementary theory (2)

I. Tick the following statements right(√) or wrong(×)

1. When working with hard strength on the bench vise, try to make the force toward the fixed vice body as much as possible. (   )

2. The plane that passes through the selected point of the cutting edge, tangent to the cutting edge and perpendicular to the base plane is the orthogonal plane. (   )

3. During cutting, the most important and most power consuming movement is the main motion. (   )

4. The displacement of the cutting tool relative to the workpiece, in the direction of the feed motion, is that back engagement of the cutting edge. (   )

5. The datum axis is the axis selected as the datum in the base axis system matching, and is indicated by "h". (   )

6. When there is any error in the blank, it can be remedied by borrowing method in marking. (   )

7. In order to make the marking clear, thicker paint should be applied to the marking part of the workpiece before marking. (   )

8. The parts must be scribed before they can be processed. (   )

9. When starting sawing, the smaller the entry angle, the better. (   )

10. The thickness of the saw blade should be selected according to the nature of the workpiece material and the width of the sawing surface. (   )

11. Hole drilling belongs to rough machining of holes. (   )

12. Commonly used cutting fluids include cooling-based water-soluble cutting fluids and lubrication-based oil-soluble cutting fluids. (   )

13. The tooth pitch of the reamer is unevenly distributed on the circumference. (   )

14. The spiral hand reamer is suitable for reaming cylindrical holes with keyways. (   )

15. When reaming, the smaller the reaming allowance, the smoother the surface after reaming. (   )

16. The diameter of the bottom hole before tapping must be greater than the minor diameter of the thread specified in the thread standard. (   )

17. Ordinary flat key connection has good alignment, convenient assembly and disassembly, and it is suitable for low-speed, high-accuracy, and occasions with variable load and impact. (   )

18. The size of the pin is usually determined on the basis that it breaks when it is overloaded by 10% to 20%. (   )

19. The transmission efficiency of the chain transmission mechanism is high, generally up

to 0.95–0.98. (　　)

20. The basic code of the bearing is composed of type code, size series code, and inner diameter code. (　　)

II. Multiple-choice questions

1. The maximum clearance in a clearance fit refers to (　　) of the upper limit size of the hole and the upper limit size of the shaft (　　).

A. sum　　B. Difference　　C. Product　　D. Quotient

2. The distance between the bracket of the grinding machine and the grinding wheel should generally be kept within (　　)mm.

A. 1　　B. 2　　C. 3　　D. 4

3. The dimensions of the part in two directions are symmetrical with the center line, and other dimensions are also marked from the center line. The marking datum of the part is (　　).

A. a plane and a center line　　B. two center lines perpendicular to each other

C. two planes perpendicular to each other　　D. two planes and a centerline

4. If the saw blade is reversely installed, its wedge angle becomes (　　).

A. unchanged　　B. bigger　　C. smaller

5. The sawing speed during sawing should be reciprocating (　　) per minute.

A. less than 20 times　　B. 20 to 30 times　　C. 20 to 40 times　　D. more than 40 times

6. The size of the rake angle of the drill bit (except the chisel edge) is related to (　　).

A. clearance angle　　B. top angle　　C. spiral angle　　D. chisel angle

7. The main purpose of adding cutting fluid during drilling is (　　).

A. lubrication effect　　B. cooling effect　　C. cleaning effect　　D. chip removal effect

8. Reaming processing belongs to the hole (　　).

A. rough machining　　B. semi-finishing machining　　C. finishing machining

9. After the reaming is completed, the reamer should (　　) to withdraw.

A. rotate forward　　B. rotate reversely

C. rotate forward and rotate reversely, both are OK.　　D. stop working

10. The diameter of the bottom hole before tapping must be (　　) the minor diameter of the thread specified in the thread standard.

A. less than　　B. greater than　　C. equal to

11. The screw on the machine that bears unidirectional force generally adopts (　　) thread.

A. sawtooth　　B. triangle　　C. circle　　D. trapezoid

12. In the V-belt transmission mechanism, movement and power are transmitted by (　　) between the belt and the pulley.

A. friction force　　B. tensile force　　C. tension force　　D. circumferential force

13. The transmission efficiency of the chain transmission mechanism is high, generally up

to ( ).

A. 0.8–0.85 B. 0.85–0.9 C. 0.9–0.95 D. 0.95–0.98

14. The code name of the straight spur gear module is ( ).

A. *z* B. *m* C. *c* D. *a*

15. In the same kind of parts, any assembly part can be installed into the parts without repair or adjustment, and the specified assembly requirements can be met. This assembly method is( ).

A. interchange method B. matching method

C. adjustment method D. repair method

16. When a tube is simmered and bent, the bending radius is generally not less than ( ) times the outer diameter of the tube.

A. 4 B.5 C.6 D.7

17. ( ) is applicable to various mechanical equipment of general medium.

A. Rectangular rubber gasket B. Oil seal gasket

C. Oil seal paper gasket D. Gasket of other materials

18. The parallelism tolerance of the center line of the main shaft taper hole of vertical drilling machine to the guide rail of column is 300mm in length, and that of the vertical and horizontal directions are both ( ).

A. 0.03 mm B. 0.04 mm C. 0.05 mm D. 0.06 mm

19. What must be done before the machine is put into trial operation? ( ).

A. Empty running test B. Load test C. Overload test D. Fatigue test

20. The working accuracy of the machine tool should be inspected by ( ) procedure.

A. first machining B. rough machining

C. semi-finishing machining D. finishing machining

Reference answer

I. Tick the following statements right(√) or wrong(×)

1-5√×√×√ 6-10 ××××√ 11-15√√×√× 16-20√××√√

II. Multiple–choice questions

1-5 BCBAC 6-10 CBBAB 11-15 AADBA 16-20 CACAD

Model test of fitter elementary theory (3)

I. Tick the following statements right(√) or wrong(×)

1. When the two jaws of the vernier caliper fit together, the zero line of the main scale and the sliding vernier scale should be aligned. (　　)

2. The cutting part of the turning tool consists of the tool bit, the main cutting edge, the auxiliary cutting edge, and the rake face and flank face. (　　)

3. The angle between the main cutting plane and the assumed working plane is the secondary deflection angle. (　　)

4. The displacement of the cutting tool relative to the workpiece in the direction of the feed motion is the back engagement of the cutting edge. (　　)

5. The dimension is a value that represents the linear length in a specific unit. (　　)

6. All parts must be marked before they can be processed. (　　)

7. The thickness of the saw blade should be selected according to the nature of the workpiece material and the width of the sawing surface. (　　)

8. For the top angle of the drill bit, drilling the hard material requires larger angle than drilling soft material. (　　)

9. The first-level maintenance of the drilling machine is mainly carried out by the operator, with the cooperation of the repairman. (　　)

10. The cutting edge on the outer circle of the cylindrical countersink is the main cutting edge, which plays a main cutting role. (　　)

11. When reaming, the smaller the reaming allowance, the smoother the surface after reaming. (　　)

12. The diameter of the bottom hole before tapping must be greater than the minor diameter of the thread specified in the thread standard. (　　)

13. The wrench is usually made of carbon structural steel or alloy structural steel. (　　)

14. Threaded connection is a fixed connection that can be assembled and disassembled. It has the advantages of simple structure, reliable connection, convenient and quick assembly and disassembly. (　　)

15. The pin, which used to determine the mutual position of the parts, is usually called the positioning pinned connection. (　　)

16. For V-belt drive, the wrap angle of the small belt pulley is generally required: $\alpha \leqslant 120°$. (　　)

17. The common damage forms of belt drive mechanism include the journal bending, loose fitting between belt pulley hole and shaft, belt pulley groove wear, belt lengthening or breaking,

cracking and so on. (   )

18. The transmission efficiency of the chain drive mechanism is high, generally up to 0.95–0.98. (   )

19. Assembly work includes pre-assembly preparation, partial assembly, final assembly, adjustment, inspection and trial operation (   )

20. Necessary cleaning, inspection, assembly and adjustment should be done in the process of mechanical installation. (   )

II. Multiple–choice questions.

1. The surface that passes through the selected point of the cutting edge, tangent to the cutting edge and perpendicular to the main movement direction is (   ).

A. front plane B. cutting plane C. base plane D. orthogonal plane

2. When marking, make the marking datum consistent with (   ).

A. design datum B. installation datum C. measurement datum D. assembly datum

3. The saw blade has a kerf, and it can make the width of the saw slot on the workpiece (   ) the thickness of the back of the saw blade.

A. less than B. equal to C. greater than

4. The selection of the section shape of the file depends on the (   ) of the workpiece.

A. shape of the file surface B. size of the file surface

C. hardness of material D. file surface accuracy

5. When grinding hardened steel parts, (   )materials are often used as grinding tools.

A. hardened steel B. low carbon steel

C. gray cast iron D. aluminum

6. The tap is composed of working part and (   ) part.

A. shank part B. calibration part C. cutting part

7. The screw on the machine that bears unidirectional force generally adopts (   ) thread.

A. sawtooth B. triangle C. circle D. trapezoid

8. In order to obtain reliable and fastened thread connection, the thread pair must be ensured to have certain (   ).

A. friction torque B. tightening torque C. pre-tightening force

9. Double nut lock mechanism belongs to (   ) anti-loosening device.

A. additional friction B. mechanical C. punch point D. bonding

10. The flat key connection transfers torque by the contact between the flat key and (   ) of the keyway.

A. upper plane B. lower plane C. both sides D. upper and lower plane

11. If the key is deformed or cut, it indicates that the key cannot withstand the transmitted torque. If the condition permits, the width of the key and the keyway or the length of the key can be increased appropriately to solve the problem. Another key can also be added to separate the

two keys at the angle of ( ) to increase the strength of the key.

A. 45° B. 90° C. 180° D. 270°

12. The length of the pin in each connecting part is approximately ( ) times the pin diameter.

A. 1-2 B. 2-3 C. 3-4 D. 4-5

13. When assembling the V-belt drive mechanism, the middle planes of the two wheels are required to coincide, so the inclination angle should not exceed ( ).

A. 10° B. 5° C. 0.1° D. 1°

14. Generally, the wrap angle of the small pulley is required as $\alpha \geqslant$ ( ).

A. 40° B. 60° C. 120° D. 180°

15. The adjustment method of tension is ( ).

A. to change the size of the belt pulley

B. to strengthen the initial tension of the belt

C. to change the center distance between the two shafts

16. The transmission efficiency of the chain drive mechanism is high, generally up to ( ).

A. 0.8-0.85 B.0.85-0.9 C.0.9-0.95 D.0.95-0.98

17. There are two methods for measuring the gear backlash: ( ).

A. coloring method and pressure fuse method

B. coloring method and indicator inspection method

C. pressure fuse method and indicator inspection method

18. The first digit of the bearing code represents ( ) of the bearing.

A. type B. width (height) series

C. diameter series D. inner diameter size

19. When assembling, the method of changing the relative position of adjustable parts in the product or selecting appropriate adjustment parts to obtain assembly accuracy is( ).

A. interchange method B. matching method

C. adjustment method D. repair method

20. The contact seal has packing seal, leather cup seal, expansion ring seal and ( ).

A. gap seal B. centrifugal seal

C. mechanical seal D. pneumatic seal

Reference answer

I. Tick the following statements right(√) or wrong(x)

1-5 √ √ × × √   6-10 × √ √ √ ×   11-15 × √ √ √ √   16-20 × √ √ × √

II. Multiple-choice questions

1-5 CACAB   6-10 AACAC   11-15 CADCC   16-20 DCACC

Model test of fitter elementary practical operation

Model test of fitter elementary practical operation (1)

I. Drawings

The drawing of concave block manufacturing is shown in following figure.

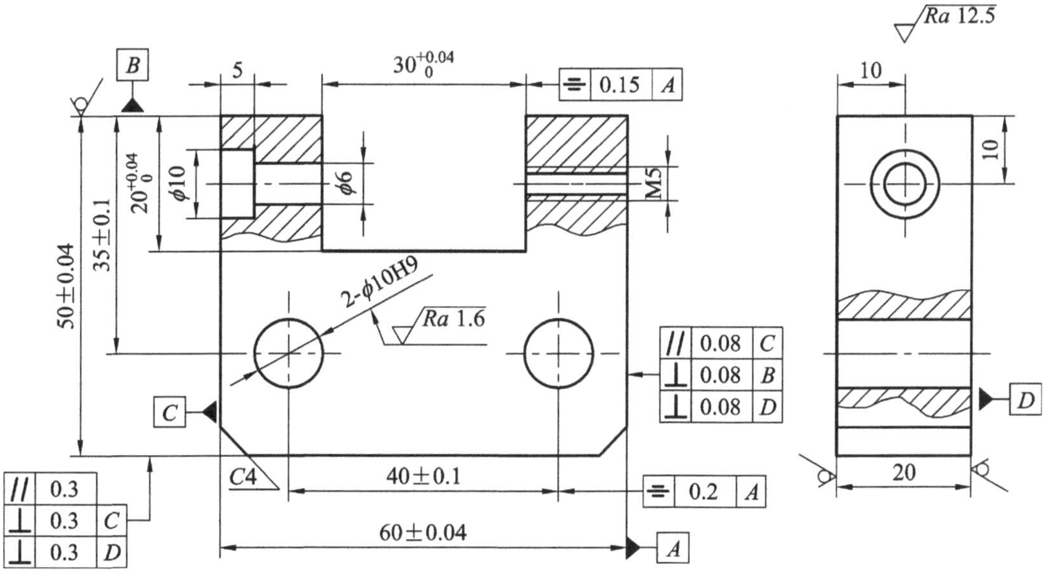

II. Assessment requirement and scoring standard

1. Assessment requirement

(1) Using sawing, filing, drilling and thread tapping methods.

(2) Surface roughness: filing *Ra*3.2 μm, drilling *Ra*6.3 μm, reaming *Ra*1.6 μm, tapping *Ra*12.5 μm.

(3) Time quota: 240 minutes.

2. Score allocation and scoring standard

No.	Assessment requirement	Score allocation	Scoring standard	Inspection result	Deduction	Score
1	(60±0.04)mm (1 place)	6	No score for out of tolerance			
2	(40±0.1)mm (1 place)	6	No score for out of tolerance			
3	(50±0.04)mm (1 place)	6	No score for out of tolerance			
4	(35±0.1)mm	6	No score for out of tolerance			
5	$20^{+0.04}_{0}$ mm	6	No score for out of tolerance			
6	$30^{+0.04}_{0}$ mm	6	No score for out of tolerance			
7	Parallelism 0.08 mm	6	No score for out of tolerance			
8	Flatness 0.3 mm	4	No score for out of tolerance			
9	Perpendicularity of *C* surface 0.3 mm (3 places)	12	No score if the requirements are not met			
10	Perpendicularity of *D* surface 0.3 mm (3 places)	12	No score for out of tolerance			
11	Symmetry 0.2 mm Symmetry 0.15 mm (1 place for each)	12	No score for out of tolerance			
12	*Ra*1.6 μm (2 places)	6	No score for one level up			
13	Saw blade	4	3 pints are deducted for each broken piece			
14	Safe and civilized production		Offenders will be deducted 3–8 points if necessary			
Field record						

III. Preparation of equipment and facilities

1. Material preparation

No.	Name	Specification	Quantity
1	Q235-A	45 mm × 40 mm × 40 mm	1 piece per person

2. Equipment preparation

No.	Name	Specification	Quantity
1	Marking plate	300 mm × 300 mm	1 piece for every 5 people
2	Bench drilling machine	Z512B	1 piece for every 5 people
3	Square box	205 mm × 205 mm × 205 mm	1 piece for every 10 people
4	Fitter's bench	2000 mm × 3000 mm	It depends on the actual number of people
5	Bench vice	100 mm or 125 mm	1 piece per person
6	Grinding machine	S3SL-250	1 piece for every 10 people

3. Preparation of machining, measuring and and cutting tools

Name	Specification	Accuracy	Quantity
Height vernier caliper	0–300 mm	0.02 mm	1 piece for every 5 people
Vernier caliper	0–150 mm	0.02 mm	1 piece for every 2 people
90° square	100 mm × 63 mm	Grade 1	1 piece for every 2 people
Knife-edge ruler	125 mm		1 piece for every 2 people
Micrometer	0–25 mm	0.01 mm	1 piece for every 2 people
	25–50 mm	0.01 mm	1 piece for every 2 people
Feeler gauge	0.02–0.5 mm		1 piece for every 2 people
Straight shank twist drill	ϕ6 mm、ϕ4.2 mm、ϕ9.8 mm、ϕ10 mm		1 piece for every 2 people
Hand cylindrical reamer	ϕ10 mm	H7	1 piece for every 2 people
Cylindrical countersink	ϕ10 mm		1 piece for every 2 people
Tap	M5		1 piece for every 2 people
Tap wrench	250 mm		1 piece for every 2 people
Flat file	300 mm(No. 1 pattern)		1 piece per person
	200 mm(No. 2 pattern)		1 piece per person
	200 mm(No. 3pattern)		1 piece per person

续表

Name	Specification	Accuracy	Quantity
Triangular file	150 mm		1 piece per person
Set file			1 piece per person
Saw bow			1 piece per person
Hammer			1 piece per person
Marking pin			1 piece per person
Sample punch			1 piece for every 2 people
Steel ruler	0-200 mm		1 piece per person
Soft jaws			1 piece per person
File brush			1 piece for every 2 people

Model test of fitter elementary practical operation (2)

I. Drawings

The drawing of limit block production is shown in following figure.

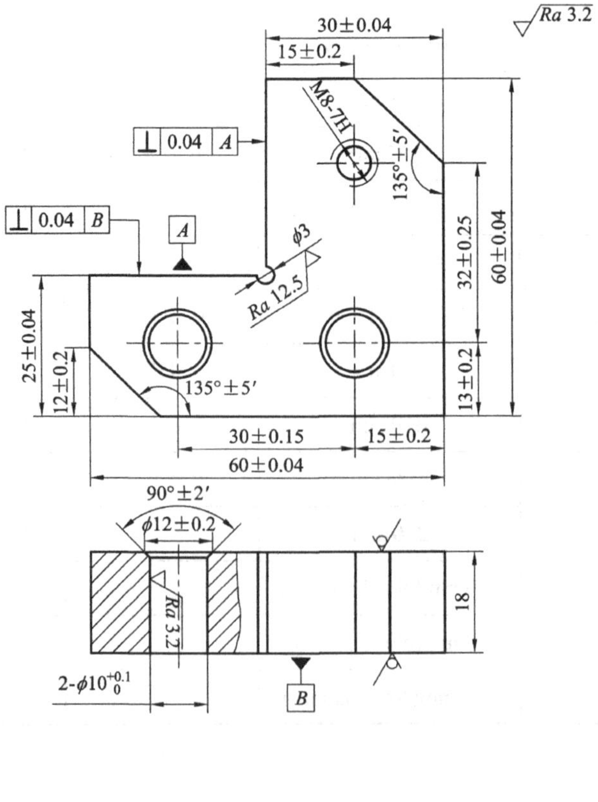

II. Assessment requirement and scoring standard

1. Assessment requirement

(1) Tolerance class: filing IT9, drilling IT11, tapping 7H.

(2) Surface roughness: filing *Ra* 3.2 μm, drilling *Ra* 3.2 μm, tapping *Ra*12.5 μm.

(3) Time quota: 240 minutes.

2. Score allocation and scoring standard

No.	Assessment requirement	Score allocation	Scoring standard	Inspection result	Deduction	Score
1	(60±0.04)mm (2 places)	10	No score for out of tolerance			
2	(30±0.04)mm	6	No score for out of tolerance			
3	(25±0.04)mm	6	No score for out of tolerance			
4	(15±0.20)mm (2 places)	5	No score for out of tolerance			
5	(12±0.20)mm	5	No score for out of tolerance			
6	135°±5′ (2 places)	6	No score for out of tolerance			
7	Perpendicularity of *A* surface 0.04 mm	12	No score for out of tolerance			
8	Perpendicularity of *B* surface 0.04 mm	12	No score for out of tolerance			
9	*Ra*3.2 μm	8	No score for one level up			
10	$2\text{-}\phi 10^{+0.1}_{0}$	4	No score for out of tolerance			
11	(32±0.25)mm	8	No score for out of tolerance			
12	*Ra*3.2 μm	4	No score for out of tolerance			
13	M8-7H	3	No score if the requirements are not met			
14	(30±0.15) mm	4	No score for out of tolerance			
15	(13±0.20) mm	4	No score for out of tolerance			
16	*Ra*12.5 μm	3	No score for one level up			
Field record						

III. Preparation of equipment and facilities

1. Material preparation

No.	Name	Specification	Quantity
1	Q235-A	65 mm × 65 mm × 18 mm	1 piece per person

2. Equipment preparation

No.	Name	Specification	Quantity
1	Marking plate	300 mm × 300 mm	1 piece for every 5 people
2	Bench drilling machine	Z512B	1 piece for every 5 people
3	Square box	205 mm × 205 mm × 205 mm	1 piece for every 10 people
4	Fitter's bench	2000 mm × 3000 mm	It depends on the actual number of people
5	Bench vice	100 mm or 125 mm	1 piece per person
6	Grinding machine	S3SL-250	1 piece for every 10 people

3. Preparation of machining, measuring and and cutting tools

Name	Specification	Accuracy	Quantity
Height vernier caliper	0–300 mm	0.02 mm	1 piece for every 5 people
Vernier caliper	0–150 mm	0.02 mm	1 piece for every 2 people
90° square	100 mm × 63 mm	Grade 1	1 piece for every 2 people
Knife-edge ruler	125 mm		1 piece for every 2 people
Micrometer	0–25 mm	0.01 mm	1 piece for every 2 people
	25–50 mm	0.01 mm	1 piece for every 2 people
Feeler gauge	0.02–0.5 mm		1 piece for every 2 people
Straight shank twist drill	ϕ6.8 mm、ϕ10 mm		1 piece for every 2 people
Cylindrical countersink	ϕ12 mm		1 piece for every 2 people
Tap	M8		1 piece for every 2 people
Tap wrench	250 mm		1 piece for every 2 people

续表

Name	Specification	Accuracy	Quantity
Flat file	300 mm(No. 1 pattern)		1piece per person
	200 mm(No. 2 pattern)		1piece per person
	200 mm(No. 3pattern)		1piece per person
Triangular file	150 mm		1piece per person
Set file			1piece per person
Saw bow			1piece per person
Hammer			1piece per person
Marking pin			1piece per person
Sample punch			1 piece for every 2 people
Steel ruler	0–200 mm		1 piece per person
Soft jaws			1 piece per person
File brush			1 piece for every 2 people

Model test of fitter elementary practical operation (3)

I. Drawings

The drawing of hexagon nut production is shown in following figure.

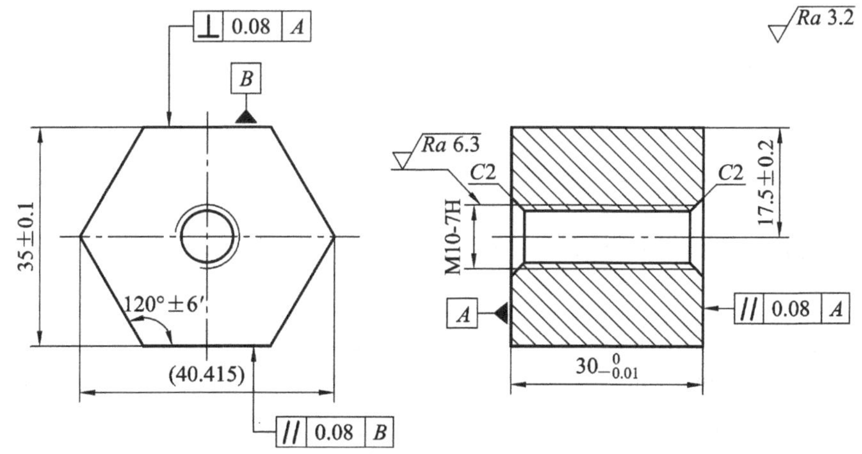

II. Assessment requirement and scoring standard

1. Assessment requirement

(1) Using sawing, filing, drilling and thread tapping methods.

(2) Surface roughness: filing *Ra*3.2 μm, drilling *Ra*6.3 μm, tapping *Ra*12.5 μm.

(3) Time quota: 240 minutes.

2. Score allocation and scoring standard

No.	Assessment requirement	Score allocation	Scoring standard	Inspection result	Deduction	Score
1	(35±0.1)mm (3 places)	6	No score for out of tolerance			
2	$30^{\ 0}_{-0.01}$mm	3	No score for out of tolerance			
3	120°±6′ (6 places)	12	No score for out of tolerance			
4	Perpendicularity 0.08 mm (6 places)	24	No score for out of tolerance			
5	Parallelism 0.08 mm (3 places)	12	No score for out of tolerance			
6	Parallelism 0.08 mm	4	No score for out of tolerance			
7	Flatness 0.05 mm (6 places)	24	No score for out of tolerance			
8	40.415 mm	3	No score for out of tolerance			
9	M10-7H	6	No score if the requirements are not met			
10	(17.5±0.2) mm	6	No score for out of tolerance			
11	Safe and civilized production		Offenders will be deducted 3-8 points if necessary			
Field record						

III. Preparation of equipment and facilities

1. Material preparation

No.	Name	Specification	Quantity
1	Q235-A	45 mm × 40 mm × 40 mm	1 piece per person

2. Equipment preparation

No.	Name	Specification	Quantity
1	Marking plate	300 mm × 300 mm	1 piece for every 5 people
2	Bench drilling machine	Z512B	1 piece for every 5 people
3	Square box	205 mm × 205 mm × 205 mm	1 piece for every 10 people
4	Fitter's bench	2000 mm × 3000 mm	It depends on the actual number of people
5	Bench vice	100 mm or 125 mm	1 person per piece
6	Grinding machine	S3SL-250	1 piece for every 10 people

3. Preparation of machining, measuring and and cutting tools

Name	Specification	Accuracy	Quantity
Height vernier caliper	0–300 mm	0.02 mm	1 piece for every 5 people
Vernier caliper	0–150 mm	0.02 mm	1 piece for every 2 people
90° square	100 mm × 63 mm	Grade 1	1 piece for every 2 people
Knife-edge ruler	125 mm		1 piece for every 2 people
Micrometer	0–25 mm	0.01 mm	1 piece for every 2 people
	25–50 mm	0.01 mm	1 piece for every 2 people
Feeler gauge	0.02–0.5 mm		1 piece for every 2 people
Straight shank twist drill	ϕ9 mm		1 piece for every 2 people
Tap	M10		1 piece for every 2 people
Tap wrench	250 mm		1 piece for every 2 people
Flat file	300 mm(No. 1 pattern)		1piece per person
	200 mm(No. 2 pattern)		1piece per person
	200 mm(No. 3pattern)		1piece per person
Triangular file	150 mm		1piece per person
Set file			1piece per person
Saw bow			1piece per person
Hammer			1piece per person
Marking pin			1piece per person
Sample punch			1 piece for every 2 people
Steel ruler	0–200 mm		1piece per person
Soft jaws			1piece per person
File brush			1 piece for every 2 people

References

[1] 徐彬. 钳工：初级[M]. 2 版 北京：机械工业出版社，2013.
[2] 杨海斌. 钳工工艺与技能训练[M]. 徐州：中国矿业大学出版社，2008.
[3] 陈刚，刘新灵. 钳工基础[M]. 北京：化学工业出版社，2014.
[4] 辛长平，左效波. 机械装配钳工基础与技能[M]. 北京：电子工业出版社，2014.
[5] 徐彬. 钳工技能鉴定考核试题库[M]. 北京：机械工业出版社，2014.
[6] 尚根宣. 钳工工艺与技能训练 [M]. 2 版. 北京：中国劳动社会保障出版社，2014.